Springer Proceedings in Mathematics & Statistics

Volume 161

Springer Proceedings in Mathematics & Statistics

This book series features volumes composed of selected contributions from workshops and conferences in all areas of current research in mathematics and statistics, including operation research and optimization. In addition to an overall evaluation of the interest, scientific quality, and timeliness of each proposal at the hands of the publisher, individual contributions are all refereed to the high quality standards of leading journals in the field. Thus, this series provides the research community with well-edited, authoritative reports on developments in the most exciting areas of mathematical and statistical research today.

More information about this series at http://www.springer.com/series/10533

Marco Castrillón López · Luis Hernández Encinas
Pedro Martínez Gadea · Mª Eugenia Rosado María
Editors

Geometry, Algebra and Applications: From Mechanics to Cryptography

In Honor of Jaime Muñoz Masqué

Springer

Editors
Marco Castrillón López
Departamento de Geometría y Topología
Universidad Complutense de Madrid
Madrid
Spain

Pedro Martínez Gadea
Institute of Fundamental Physics
Spanish National Research Council (CSIC)
Madrid
Spain

Luis Hernández Encinas
Physical and Information Technology
Spanish National Research Council (CSIC)
Madrid
Spain

Mª Eugenia Rosado María
Departamento de Matemática Aplicada
Universidad Politécnica de Madrid
Madrid
Spain

ISSN 2194-1009 ISSN 2194-1017 (electronic)
Springer Proceedings in Mathematics & Statistics
ISBN 978-3-319-32084-7 ISBN 978-3-319-32085-4 (eBook)
DOI 10.1007/978-3-319-32085-4

Library of Congress Control Number: 2016939063

Mathematics Subject Classification (2010): 53-06, 11-06, 16-06

Printed on acid-free paper

This Springer imprint is published by Springer Nature
The registered company is Springer International Publishing AG Switzerland

A las aladas almas de las rosas
del almendro de nata te requiero,
que tenemos que hablar de muchas cosas,
compañero del alma, compañero.

Elegía a Ramón Sijé (1936)
Miguel Hernández

Contents

About the Editors

Prof. Marco Castrillón López read mathematics and physics in Madrid. He is currently Professor Titular at the Universidad Complutense de Madrid, where he also received his Ph.D. in Mathematics. He was a postdoc at École Polytechnique Fédérale de Lausanne (Switzerland), and Faculty Visitor at Caltech (Pasadena, USA), PIMS (Vancouver, Canada), Imperial College (London, UK), TATA Institute (Mumbai, India) and PUC (Rio de Janeiro, Brazil). His research work mainly focuses on geometric variational calculus, gauge theories and Riemannian geometry with applications to relativity, classical field theories and other topics in theoretical physics. His has over 50 publications and books to his name.

Luis Hernández Encinas graduated in mathematics at the University of Salamanca (Spain) in 1980, and received his Ph.D. in Mathematics from the same university in 1992. He is Researcher at the Department of Information Processing and Cryptography (TIC) in the Institute of Physical and Information Technologies (ITEFI), Spanish National Research Council (CSIC) in Madrid. He has participated in more than 30 research projects. He is author of nine books, nine patents, more than 150 papers, and over 100 contributions to workshops and conferences. He has also supervised several doctoral theses. His current research interests include cryptography and cryptanalysis of public key cryptosystems, digital signature schemes, authentication and identification protocols, crypto-biometry, side channel attacks and number theory problems.

Prof. Pedro Martínez Gadea taught at the Universities of Santiago de Compostela and Valladolid in Spain. He is now scientific Researcher at the Instituto de Física Fundamental, CSIC, Madrid, Spain. He has published almost 70 research papers on several topics of differential geometry and algebraic topology. He has also been advisor for four Ph.D. theses. His current interests include differential geometry, more specifically in homogeneous spin Riemannian manifolds and Ricci-flat invariant Kähler structures.

Mª Eugenia Rosado María graduated in mathematics at the University Complutense de Madrid, and obtained her Ph.D. in Mathematics from the same university. She previously taught at the Universities Autónoma de Madrid, Spain and currently teaches at the Universidad Politécnica de Madrid. She has published around 20 papers on several topics in differential geometry. Her research interests include geometrical variational calculus, geometric methods in differential equations, differential invariants and other topics in differential geometry.

Introduction

This volume presents a collection of articles to honour Prof. Jaime Muñoz Masqué on the occasion of his 65th birthday. Jaime was born on 20 September 1950 in Sabadell, Barcelona (Spain), to his parents, Manuel and Rosenda. He attended high school in his home village, initially showing an inclination towards literature and poetry, which he subsequently combined with a strong interest in mathematical problems. He devoted more time to the latter; for instance, he spent a summer of his adolescence exploring the intriguing question of the unsolvability of the equations defined by fifth degree polynomials. This clearly indicated that his destiny was to study mathematics, which he did at the University of Barcelona.

During his studies at the university, Jaime made acquaintance of two important persons: First, María Sicilia, his future wife, who was also studying mathematics, and second, Pedro Luis García Pérez, with whom he decided to do his Ph.D. As Prof. García held a post in the University of Salamanca, the newly established family moved to this city after both María and Jaime had completed their studies in 1973. Jaime won his position as High School Professor (Catedrático) in 1975, working first in Zamora and then in Alba de Tormes (Salamanca). Jaime helped some of his colleagues at the High School María de Molina to prepare for their national-level exams in order to obtain permanent positions. They all remember these years with affection.

In the meantime, Jaime had also begun to lecture at the University of Salamanca. In 1983, Jaime defended his doctoral thesis at that university, entitled *Hamilton-Cartan Theory for higher-order variational problems on fibered manifolds* (*Teoría de Hamilton-Cartan para los problemas variacionales de orden superior sobre variedades fibradas*). He also began his fruitful scientific career with the publication of his papers. In his first article, *Higher-order structure forms and infinitesimal contact transformations* (*Formes de structure et transformations infinitésimales de contact d'ordre supérieur*, CR Acad Sci Paris Sér I Math 1984; 298, no. 8:185–8), he formalized the geometry behind the natural lift of vector fields from a bundle to its jet extension for arbitrary degree. This tool was essential for his work on higher-order variational calculus, the topic of his thesis, on which he became a world expert. At the same time, his family grew as María and Jaime had their three children: Ana, Joaquín and Teresa.

From 1984 to 1989, Jaime was first Assistant Professor and then Associate Professor at the University of Salamanca, where he continued his scientific work, mainly in the fields of differential geometry and algebra. In 1989 he was appointed as Researcher at CSIC (Spanish National Research Council) and the family moved to Madrid. While working in his new position he added cryptography to the list of his interests, and joined research projects in this field. He continued collaborating in CSIC and has carried out research on these topics until now.

In parallel with his scientific work, Jaime delivered courses in different universities where he showed his rare talent of explaining complex mathematics in a clean, simple and rigorous language. In association with this academic work, he has been the advisor for nine doctoral theses (Marco Castrillón López, Raúl Durán Díaz, Víctor Fernández Mateos, Roberto Ferreiro, Ángel Martín del Rey, Alberto Peinado Domínguez, Luis Pozo Coronado, Eugenia Rosado María, Antonio Valdés) covering a varied collection of topics in geometry and algebra, from variational calculus, Riemannian geometry and theory of invariants to cryptography. We have borne in mind this versatility for choosing the title of this volume, which offers an indication of Jaime's vast knowledge and wide-ranging scientific works. In this respect, the database of the Mathematical Reviews of the AMS includes as many as 162 contributions from Jaime, including both books and articles, on which he has worked with 39 collaborators.

The general consensus among the people who work with Jaime is that he is not only a hard worker but also possesses a very broad knowledge of mathematics and physics (as well as poetry and philosophy!) and an incredible capability to tackle problems in very different areas in an interdisciplinary atmosphere. We all enjoy his warm personality, the conversations with him over a cup of coffee and especially his generosity, in all senses of the word. Jaime is a person who loves mathematics and with whom one feels that excitement which accompanies the search for a solution or the thrilling experience of finding those hidden mathematical gems accessible only to a select group—a group of which Jaime is undoubtedly a member.

Marco Castrillón López
Luis Hernández Encinas
Pedro Martínez Gadea
Mª Eugenia Rosado María

A Survey on Homogeneous Structures on the Classical Hyperbolic Spaces

Wafaa Batat, P.M. Gadea and José A. Oubiña

Dedicated to our colleague and friend Jaime Muñoz Masqué, a good mathematician, with affection and admiration, on the occasion of his 65th birthday

Abstract This is a survey on homogeneous Riemannian, Kähler or quaternionic Kähler structures on the real, complex or quaternionic hyperbolic spaces $\mathbb{R}\mathrm{H}(n)$, $\mathbb{C}\mathrm{H}(n)$ and $\mathbb{H}\mathrm{H}(n)$, respectively.

Keywords Homogeneous Riemannian structures · Classical hyperbolic spaces

1 Introduction

Real, complex and quaternionic hyperbolic spaces and the Cayley hyperbolic plane are known to be important spaces and have been and are subject of much research. Two general references are Chen and Greenberg [10] and Ratcliffe [22].

On the other hand, homogeneous Riemannian structures were introduced by Ambrose and Singer [3], and further studied in depth by Tricerri and Vanhecke

W. Batat
Département de Mathématiques et Informatique, Ecole Normale Supérieure d'Enseignement Technologique d'Oran, B.P. 1523, El M'Naouar, Oran, Algeria
e-mail: batatwafa@yahoo.fr

P.M. Gadea (✉)
Instituto de Física Fundamental, CSIC, Serrano 113-bis, 28006 Madrid, Spain
e-mail: p.m.gadea@csic.es

J.A. Oubiña
Facultade de Matemáticas, Departamento de Xeometría e Topoloxía, Universidade de Santiago de Compostela, A Coruna, Spain
e-mail: ja.oubina@usc.es

M. Castrillón López et al. (eds.), *Geometry, Algebra and Applications: From Mechanics to Cryptography*, Springer Proceedings in Mathematics & Statistics 161, DOI 10.1007/978-3-319-32085-4_1

(see for instance [25]) and then by other authors. There exist three basic geometric types, $\mathcal{S}_1$, $\mathcal{S}_2$, $\mathcal{S}_3$. Later, homogeneous Kähler structures were defined and studied by Abbena and Garbiero in [1] and then by several authors. This time there are four basic types, $\mathcal{K}_1, \ldots, \mathcal{K}_4$. Further, homogeneous quaternionic Kähler structures were introduced by Fino [11], who moreover gave a Lie-theoretical description of the five basic types, $\mathcal{QK}_1, \ldots, \mathcal{QK}_5$, and then studied by several authors. (In the sequel we shall denote $\mathcal{S}_i \oplus \mathcal{S}_j$ simply by $\mathcal{S}_{ij}$; $\mathcal{K}_i \oplus \mathcal{K}_j$ by $\mathcal{K}_{ij}$; $\mathcal{QK}_i \oplus \mathcal{QK}_j$ by $\mathcal{QK}_{ij}$, and so on.)

Homogeneous Riemannian structures have found some useful applications. Two of them are: The characterization of $\mathbb{R}\mathrm{H}(n)$, $\mathbb{C}\mathrm{H}(n)$ and $\mathbb{H}\mathrm{H}(n)$ by such structures and the characterization of the homogeneous spin Riemannian manifolds whose Dirac operator is like that on a Riemannian symmetric spin space (see [15]). In our opinion, Tricerri's and Vanhecke's classification of geometric types is so natural, that more nice applications are to be expected.

The present survey is on the characterization of each of the classical hyperbolic spaces by linear homogeneous structures and on the geometric types of homogeneous structures on them. Recall that the characterization of $\mathbb{R}\mathrm{H}(n)$ by homogeneous Riemannian structures of type $\mathcal{S}_1$ was given by Tricerri and Vanhecke in [25], that of $\mathbb{C}\mathrm{H}(n)$ in terms of homogeneous Kähler structures of type $\mathcal{K}_{24}$ was obtained in [16], and that of $\mathbb{H}\mathrm{H}(n)$ by homogeneous quaternionic Kähler structures of type $\mathcal{QK}_{123}$ with nonzero projection to $\mathcal{QK}_3$ (actually, of type $\mathcal{QK}_3$) was given in [7].

The vector spaces $\mathcal{S}_1$, $\mathcal{K}_{24}$ and $\mathcal{QK}_{123}$ have dimension growing linearly according to the dimension of the homogeneous manifold admitting some homogeneous structure in each of them, that is, hyperbolic spaces. For this reason, these structures are sometimes called of linear type. However, this is not the unique type that hyperbolic spaces admit.

As for the contents, we recall in Sect. 2 some definitions on homogeneous Riemannian, Kähler and quaternionic Kähler structures, and recall the classification of geometric types for each of the three cases.

In Sect. 3 we give some results on the types of homogeneous structures that $\mathbb{R}\mathrm{H}(n)$, $\mathbb{C}\mathrm{H}(n)$ or $\mathbb{H}\mathrm{H}(n)$ admit.

2 Homogeneous Riemannian, Kähler or Quaternionic Kähler Structures

2.1 Homogeneous Riemannian Structures

A homogeneous structure on a Riemannian manifold (M, g) is a tensor field S of type $(1, 2)$ satisfying

$$\widetilde{\nabla} g = 0, \quad \widetilde{\nabla} R = 0, \quad \widetilde{\nabla} S = 0, \tag{1}$$

where $\widetilde{\nabla}$ is (see [25]) the connection determined by $\widetilde{\nabla} = \nabla - S$, ∇ being the Levi–Civita connection of g. The condition $\widetilde{\nabla} g = 0$ is equivalent to $S_{XYZ} = -S_{XZY}$, where $S_{XYZ} = g(S_X Y, Z)$.

Ambrose and Singer [3] gave the following characterization of homogeneous Riemannian manifolds: *A connected, simply connected and complete Riemannian manifold (M, g) is homogeneous if and only if it admits a homogeneous structure S.*

Let V be a real vector space endowed with an inner product $\langle\cdot,\cdot\rangle$, which is the model for each tangent space T_pM, $p \in M$, of a (homogeneous) Riemannian manifold. Consider the vector space $\mathscr{S}(V)$ of tensors of type (0, 3) on $(V, \langle\cdot,\cdot\rangle)$ satisfying the same algebraic symmetry that a homogeneous Riemannian structure S, that is, $\mathscr{S}(V) = \{S \in \otimes^3 V^* : S_{XYZ} = -S_{XZY}, X, Y, Z \in V\}$.

Tricerri and Vanhecke studied the decomposition of $\mathscr{S}(V)$ into invariant and irreducible subspaces $\mathscr{S}_i(V)$, $i = 1, 2, 3$, under the action of the orthogonal group $O(n)$ given by $(aS)_{XYZ} = S_{a^{-1}X\, a^{-1}Y\, a^{-1}Z}$, $a \in O(n)$. The inner product on V induces in a natural way an inner product on $\mathscr{S}(V)$, given by $\langle S, S'\rangle = \sum_{i,j,k=1}^{n} S_{e_ie_je_k} S'_{e_ie_je_k}$, where $\{e_i\}$ is an orthonormal basis of V. Let $c_{12}(S)(Z) = \sum_{i=1}^{n} S_{e_ie_iZ}$, $Z \in V$.

From the theory of representations of the orthogonal group (cf. [26, pp. 153–159]) it follows that $\mathscr{S}(V)$ decomposes into the orthogonal direct sum of three invariant and irreducible subspaces under the action of $O(n)$. Specifically, the subspace of c_{12}-traceless tensors of the subspace $\mathscr{Y}$ of $\otimes^3 V^*$ corresponding to the nonstandard Young symmetrizer id $+ (12) - (23) - (132)$, the n -dimensional subspace of tensors corresponding to the above c_{12}-trace, and the subspace $\wedge^3 V^*$. Then, one has

Theorem 1 ([25]) *If* $\dim V \geqslant 3$, *then $\mathscr{S}(V)$ decomposes into the orthogonal direct sum of subspaces which are invariant and irreducible under the action of $O(n)$, $\mathscr{S}(V) = \mathscr{S}_1(V) \oplus \mathscr{S}_2(V) \oplus \mathscr{S}_3(V)$, where*

$$\begin{aligned}
\mathscr{S}_1(V) &= \{S \in \mathscr{S}(V) : S_{XYZ} = \langle X, Y\rangle\theta(Z) - \langle X, Z\rangle\theta(Y),\ \theta \in V^*\},\\
\mathscr{S}_2(V) &= \{S \in \mathscr{S}(V) : \mathfrak{S}_{XYZ} S_{XYZ} = 0,\ c_{12}(S) = 0\},\\
\mathscr{S}_3(V) &= \{S \in \mathscr{S}(V) : S_{XYZ} + S_{YXZ} = 0\},
\end{aligned}$$

with dimensions n, $\frac{1}{3}n(n^2-4)$, $\frac{1}{6}n(n-1)(n-2)$, respectively. If $\dim V = 2$ *then $\mathscr{S}(V) = \mathscr{S}_1(V)$.*

We say that the homogeneous Riemannian structure S on (M, g) is of *type* $\{0\}$, $\mathscr{S}_i$ $(i = 1, 2, 3)$, $\mathscr{S}_{ij}$ $(1 \leqslant i < j \leqslant 3)$, or $\mathscr{S}_{123}$ if, for each point $p \in M$, $S(p) \in \mathscr{S}(T_pM)$ belongs to $\{0\}$, $\mathscr{S}_i(T_pM)$, $\mathscr{S}_{ij}(T_pM)$ or $\mathscr{S}_{123}(T_pM)$, respectively.

The similar terminology and notation will be used for the homogeneous Kähler (Sect. 2.2) and homogeneous quaternionic Kähler (Sect. 2.3) geometric types, that is, for the different types obtained from the basic types $\mathscr{K}_i$ $(i = 1, \dots, 4)$ and $\mathscr{QK}_i$ $(i = 1, \dots, 5)$, respectively.

2.2 Homogeneous Kähler Structures

An almost Hermitian manifold (M, g, J) is called a homogeneous almost Hermitian manifold if there exists a Lie group of almost complex isometries acting transitively and effectively on M. In [24], Sekigawa proved that a simply connected and complete almost Hermitian manifold (M, g, J) is homogeneous if and only if it admits a tensor field S of type $(1, 2)$ satisfying the Ambrose–Singer equations (1) and $\widetilde{\nabla} J = 0$. Such a tensor field S is called a homogeneous almost Hermitian structure (or a homogeneous Kähler structure if (M, g, J) is Kähler). Moreover, a homogeneous Riemannian structure on a Kähler manifold (M, g, J) is a homogeneous Kähler structure if and only if $S_{ZXY} = S_{Z\,JX\,JY}$ for all vector fields X, Y, Z on M.

The classification of homogeneous Kähler structures was obtained by Abbena and Garbiero. We recall here their result: Let V be a $2n$-dimensional real vector space (which is the model for the tangent space at any point of a manifold equipped with a Kähler homogeneous structure) endowed with a complex structure J and a Hermitian inner product $\langle\ ,\ \rangle$, that is, $J^2 = -I$, $\langle JX, JY\rangle = \langle X, Y\rangle, X, Y \in V$, where I denotes the identity isomorphism of V.

Denoting complexifications by a superscript c, we now consider the decompositions in ($\pm$i)-eigenspaces $V^c = V^{1,0} \oplus V^{0,1}$ and $V^{*c} = \lambda^{1,0} \oplus \lambda^{0,1}$, with respect to the complexified J^c of the complex structure J. In Salamon's notation [23], let $\lambda^{p,q}$ denote the space of forms of type (p, q), which is isomorphic to $\Lambda^p\lambda^{1,0} \otimes \Lambda^q\lambda^{0,1}$. We can decompose the space $\mathscr{S}(V)^c = \{S \in \otimes^3 V : S_{XYZ} = -S_{XZY}\}, X, Y, Z \in V^c$, into two subspaces invariant under the action of $U(n)$. One summand (that is, $\mathscr{S}(V)^c_- = V^{*c} \otimes (\lambda^{2,0} \oplus \lambda^{0,2})$) is related to homogeneous almost Hermitian structures. The other summand is

$$\mathscr{S}(V)^c_+ = V^{*c} \otimes \lambda^{1,1} \cong \{S \in \otimes^3 V : S_{XYZ} = -S_{XZY} = S_{XJ^cYJ^cZ}\},$$

$X, Y, Z \in V^c$, which is the complexified of Abbena–Garbiero's space $\mathscr{S}(V)_+$ (see [1]). The space $S(V)_+$ decomposes ([12, (2.1)]) into four subspaces invariant and irreducible under the action of $U(n)$. The sum of the first and second subspaces corresponds with the irreducible complex representation of $U(n)$ of the highest weight $(1, 1, 0, \dots, 0, -1)$. The related real tensors of trace zero and those corresponding to that trace give rise to the first and second types in Theorem 2 below. Similarly, the sum of the third and four subspaces in that theorem, corresponds to the irreducible complex representation of $U(n)$ of the highest weight $(2, 0, \dots, 0, -1)$. Taking traceless real tensors one gets the third subspace and the fourth one comes from that trace. We recall that Abbena and Garbiero [1, Theorem 4.4] proved the invariance and irreducibility by using quadratic invariants. In [12], Young diagrams and symmetrizers are used instead.

The standard representation of $U(n)$ on V induces a representation of $U(n)$ on $\mathscr{S}(V)_+$ given by $(A(S))_{XYZ} = S_{A^{-1}XA^{-1}YA^{-1}Z}, A \in U(n)$. Moreover, the scalar product in V induces in a natural way the scalar product in $\mathscr{S}(V)$ given by $\langle S, S'\rangle = \sum_{i,j,k=1}^{2n} S_{e_ie_je_k} S'_{e_ie_je_k}$, for any orthonormal basis $\{e_1, \dots, e_{2n}\}$ of V. The expression of the tensors in each basic geometric type was given by Abbena and Garbiero and is as follows.

Theorem 2 ([1]) *If* $\dim V \geqslant 6$, $\mathscr{S}(V)_+$ *decomposes into the orthogonal direct sum of the following subspaces invariant and irreducible under the action of the group* $U(n)$*:*

$$\begin{aligned}
\mathscr{K}_1 =& \ \{S \in \mathscr{S}(V) : S_{XYZ} = \tfrac{1}{2}(S_{YZX} + S_{ZXY} + S_{JYJZX} + S_{JZXJY}),\ c_{12}(S) = 0\},\\
\mathscr{K}_2 =& \ \{S \in \mathscr{S}(V) : S_{XYZ} = \langle X, Y\rangle\theta_1(Z) - \langle X, Z\rangle\theta_1(Y) + \langle X, JY\rangle\theta_1(JZ)\\
&\qquad - \langle X, JZ\rangle\theta_1(JY) - 2\langle JY, Z\rangle\theta_1(JX),\ \theta_1 \in V^*\},\\
\mathscr{K}_3 =& \ \{S \in \mathscr{S}(V) : S_{XYZ} = -\tfrac{1}{2}(S_{YZX} + S_{ZXY} + S_{JYJZX} + S_{JZXJY}),\ c_{12}(S) = 0\},\\
\mathscr{K}_4 =& \ \{S \in \mathscr{S}(V) : S_{XYZ} = \langle X, Y\rangle\theta_2(Z) - \langle X, Z\rangle\theta_2(Y) + \langle X, JY\rangle\theta_2(JZ)\\
&\qquad - \langle X, JZ\rangle\theta_2(JY) + 2\langle JY, Z\rangle\theta_2(JX),\ \theta_2 \in V^*\},
\end{aligned}$$

$X, Y, Z \in V$, *where* c_{12} *is defined by* $c_{12}(S)(X) = \sum_{i=1}^{2n} S_{e_i e_i X}$, *for any* $X \in V$ *and* $\{e_1, \ldots, e_{2n}\}$ *being an orthonormal basis of* V*;* $\theta_1(X) = (1/(2(n-1)))c_{12}(S)(X)$ *and* $\theta_2(X) = (1/(2(n+1)))c_{12}(S)(X)$, $X \in V$. *The dimensions are* $n(n+1)(n-2)$, $2n$, $n(n-1)(n+2)$ *and* $2n$, *respectively. If* $\dim V = 4$, *then* $\mathscr{S}(V)_+ = \mathscr{K}_2 \oplus \mathscr{K}_3 \oplus \mathscr{K}_4$. *If* $\dim V = 2$, *then* $\mathscr{S}(V)_+ = \mathscr{K}_4$.

2.3 Homogeneous Quaternionic Kähler Structures

Let (M, g, υ^3) be an almost quaternion-Hermitian $4n$-manifold, υ^3 being the structure subbundle of the bundle of $(1, 1)$ tensors on M and let ∇ denote the Levi–Civita connection. The manifold is said to be quaternion-Kähler if one has locally (cf. Ishihara [19]) that

$$\nabla_X J_1 = \tau^3(X)J_2 - \tau^2(X)J_3, \quad \text{etc.,} \tag{2}$$

for certain differential 1-forms τ^1, τ^2, τ^3. Here and in the sequel we write "etc." to indicate the similar formulas obtained by cyclic permutation of (123). The holonomy group is contained in $Sp(n)Sp(1)$. A quaternion-Kähler manifold (M, g, υ^3) is said to be a *homogeneous quaternion-Kähler manifold* if it admits a transitive group of isometries (cf. Alekseevsky and Cortés [2, p.218] and [7, Remark 2.2]). A connected, simply connected and complete quaternion-Kähler manifold (M, g, υ^3) is homogeneous if and only if it admits a *homogeneous quaternionic Kähler structure*, that is, a $(1, 2)$ tensor field S satisfying the Ambrose–Singer equations (1) and equations

$$\widetilde{\nabla}_X J_1 = \tilde{\tau}^3(X)J_2 - \tilde{\tau}^2(X)J_3, \quad \text{etc.,} \tag{3}$$

for three differential 1-forms $\tilde{\tau}^1, \tilde{\tau}^2, \tilde{\tau}^3$. Let $\theta^a = \tau^a - \tilde{\tau}^a$, $a = 1, 2, 3$. Then, from formulas (2) and (3) we have that

$$S_{XJ_1YJ_1Z} - S_{XYZ} = \theta^3(X)g(J_2Y, J_1Z) - \theta^2(X)g(J_3Y, J_1Z), \quad \text{etc.,}$$

which, together with the condition $S_{XYZ} = -S_{XZY}$, are the algebraic symmetries satisfied by a homogeneous quaternionic Kähler structure S.

Denote by E the standard representation of $Sp(n)$ on $\mathbb{C}^{2n}$, by S^rE the rth-symmetric power of E (so that $S^2E \cong \mathfrak{sp}(n) \otimes \mathbb{C}$), by K the irreducible $Sp(n)$-module of the highest weight $(2, 1, 0, \dots, 0)$ in $E \otimes S^2E = S^3E \oplus K \oplus E$, and by H the standard representation of $Sp(1) \cong SU(2)$ on $\mathbb{C}^2$, so that $S^2H \cong \mathfrak{sp}(1) \otimes \mathbb{C}$ and S^3H is the 4-dimensional irreducible representation of $Sp(1)$.

Let $\mathscr{S}(V)_+$ denote the set of homogeneous quaternionic Kähler structures. The geometric types were classified from a representation-theoretic point of view as follows.

Theorem 3 (Fino [11, Lemma 5.1])

$$\mathscr{S}(V)_+ = [EH] \otimes (\mathfrak{sp}(1) \oplus \mathfrak{sp}(n)) \cong [EH] \oplus [ES^3H] \oplus [EH] \oplus [S^3EH] \oplus [KH].$$

Here, $[V]$ denotes the real representation whose complexification is V and the tensor products signs are omitted, that is, one writes EH instead of $E \otimes H$, and so on.

The standard representation $[EH]$ of $Sp(n)Sp(1)$ on V induces a representation of $Sp(n)Sp(1)$ on $\mathscr{S}(V)_+$ given by $(A(S))_{XYZ} = S_{A^{-1}XA^{-1}YA^{-1}Z}$, $A \in Sp(n)Sp(1)$. Moreover, the scalar product in V induces in a natural way the scalar product in $\mathscr{S}(V)_+$ given by $\langle S, S' \rangle = \sum_{i,j,k=1}^{4n} S_{e_ie_je_k} S'_{e_ie_je_k}$, for any orthonormal basis $\{e_1, \dots, e_{4n}\}$ of V. The classification of homogeneous quaternionic Kähler structures in terms of real tensors was given in [7], as we now recall (except for the explanation of a few notations).

Theorem 4 ([7, Theorem 1.1]) *If $n \geqslant 2$, then $\mathscr{V}$ decomposes into the orthogonal direct sum of the following subspaces invariant and irreducible under the action of $Sp(n)Sp(1)$:*

$$\mathcal{QK}_1 = \{\Theta \in \widetilde{\mathscr{V}} : \Theta_{XYZ} = \textstyle\sum_{a=1}^3 \theta(J_aX)\langle J_aY, Z\rangle,\ \theta \in V^*\},$$

$$\mathcal{QK}_2 = \{\Theta \in \widetilde{\mathscr{V}} : \Theta_{XYZ} = \textstyle\sum_{a=1}^3 \theta^a(X)\langle J_aY, Z\rangle,\ = \sum_{a=1}^3 \theta^a \circ J_a = 0,\ \theta^1, \theta^2, \theta^3 \in V^*\},$$

$$\mathcal{QK}_3 = \{S \in \widehat{\mathscr{V}} : S_{XYZ} = \langle X, Y\rangle\theta(Z) - \langle X, Z\rangle\theta(Y)$$
$$+ \textstyle\sum_{a=1}^3 (\langle X, J_aY\rangle\theta(J_aZ) - \langle X, J_aZ\rangle\theta(J_aY)), \theta \in V^*\},$$

$$\mathcal{QK}_4 = \{S \in \widehat{\mathscr{V}} : S_{XYZ} = \tfrac{1}{6}(\mathfrak{S}_{XYZ}S_{XYZ} + \textstyle\sum_{a=1}^3 \mathfrak{S}_{XJ_aYJ_aZ}S_{XJ_aYJ_aZ}), c_{12}(S) = 0\},$$

$$\mathcal{QK}_5 = \{S \in \widehat{\mathscr{V}} : \mathfrak{S}_{XYZ}S_{XYZ} = 0\},$$

with dimensions $4n$, $8n$, $4n$, $\frac{4}{3}n(n+1)(2n+1)$, $\frac{16}{3}n(n^2-1)$, *respectively.*

3 Types of Homogeneous Structures on $\mathbb{R}\mathrm{H}(n)$, $\mathbb{C}\mathrm{H}(n)$ or $\mathbb{H}\mathrm{H}(n)$

The usual homogeneous description of each hyperbolic space is as a rank-one noncompact Riemannian symmetric space, that is, as $\mathbb{R}\mathrm{H}(n) = SO(n,1)/O(n)$, $\mathbb{C}\mathrm{H}(n) = SU(n,1)/S(U(n) \times U(1))$ and $\mathbb{H}\mathrm{H}(n) = Sp(n,1)/(Sp(n) \times Sp(1))$, respectively. Then the corresponding homogeneous tensor S vanish.

We have the next result.

Proposition 1 **(i)** ([25, Theorem 5.2]) *A connected, simply connected and complete Riemannian manifold of dimension $n \geqslant 2$ admits a nontrivial homogeneous structure $S \in \mathscr{S}_1$ if and only if it is isometric to* $\mathbb{R}\mathrm{H}(n)$.

(ii) ([16, Theorem 1.1]) *A connected, simply connected and complete irreducible Kähler manifold of dimension $2n \geqslant 4$ admits a nontrivial homogeneous Kähler structure $S \in \mathscr{K}_{24}$ if and only if it is holomorphically isometric to* $\mathbb{C}\mathrm{H}(n)$.

(iii) ([7, Theorem 1.1]) *A connected, simply connected and complete quaternionic Kähler manifold of dimension $4n \geqslant 8$ admits a nontrivial homogeneous quaternionic Kähler structure $S \in \mathscr{QK}_{123}$ if and only if it is isometric to $\mathbb{H}\mathrm{H}(n)$. In this case, the homogeneous structure is necessarily of type $\mathscr{QK}_3$.*

Recall (Heintze [18, Theorem 4]), that a connected homogeneous Kähler $2n$-manifold of negative curvature is holomorphically isometric to $\mathbb{C}\mathrm{H}(n)$. Hence from Proposition 1, (ii), it follows the next

Corollary 1 *Any connected homogeneous Kähler manifold of real dimension $2n \geqslant 4$ and negative curvature admits a Kähler homogeneous structure $S \in \mathscr{K}_{24}$.*

However, hyperbolic spaces admit more types of homogeneous structures. We first recall

Proposition 2 **(i)** ([8, Theorem 3.1]) *The connected groups acting transitively on $\mathbb{R}\mathrm{H}(n)$ are the full isometry group $SO(n,1)$ and the groups $G = F_r N$, where N is the nilpotent factor in the Iwasawa decomposition of $SO(n,1)$ and F_r is a connected closed subgroup of $SO(n-1)\mathbb{R}$ with nontrivial projection to $\mathbb{R}$.*

(ii) ([8, Theorem 4.1]) *The connected groups acting transitively on $\mathbb{C}\mathrm{H}(n)$ are the full isometry group $SU(n,1)$ and the groups $G = F_r N$, where N is the nilpotent factor in the Iwasawa decomposition KAN of $SU(n,1)$ and F_r is a connected closed subgroup of $S(U(n-1)U(1))\mathbb{R}$ with nontrivial projection to $\mathbb{R}$.*

(iii) ([7, Theorem 5.2]) *The connected groups acting transitively on $\mathbb{H}\mathrm{H}(n)$ are the full isometry group $Sp(n,1)$ and the groups $G = F_r N$, where N is the nilpotent factor in the Iwasawa decomposition KAN of $Sp(n,1)$ and F_r is a connected closed subgroup of $Sp(n-1)Sp(1)\mathbb{R}$ with nontrivial projection to $\mathbb{R}$.*

The simplest choice is $F_r = A$, giving the description of $\mathbb{R}\mathrm{H}(n)$, $\mathbb{C}\mathrm{H}(n)$ or $\mathbb{H}\mathrm{H}(n)$ as the solvable group AN, and one has

Proposition 3 **(i)** ([8, Subsection 3.1]) *Any homogeneous Riemannian structure on* $\mathbb{R}\mathrm{H}(n) \equiv AN$ *with trivial holonomy lies in the class* $\mathscr{S}_1$.

(ii) ([8, Proposition 4.2]) *Any homogeneous Kähler structure on* $\mathbb{C}\mathrm{H}(n) \equiv AN$ *with trivial holonomy lies in the class* $\mathscr{K}_{234}$.

(iii) ([7, Proposition 5.3]) *Any homogeneous quaternionic Kähler structure on* $\mathbb{H}\mathrm{H}(n) \equiv AN$ *with trivial holonomy lies in the class* $\mathscr{QK}_{134}$.

For structures of linear type one has

Proposition 4 ([25, p. 55], [8, Subsection 3.1]) **(i)** *The homogeneous Riemannian structures of linear type on* $\mathbb{R}\mathrm{H}(n)$ *can be realized by the homogeneous model AN, where AN stands for the solvable part of the Iwasawa decomposition of the full isometry group* $SO(n, 1)$.

(ii) ([8, Theorem 4.4]) *The homogeneous Kähler structures of linear type on* $\mathbb{C}\mathrm{H}(n)$ *can be realized by the homogeneous model* $U(1)AN/U(1)$*, where AN stands for the solvable part of the Iwasawa decomposition of the full isometry group* $SU(n, 1)$.

(iii) ([7, Theorem 5.4]) *The homogeneous quaternionic Kähler structures of linear type on* $\mathbb{H}\mathrm{H}(n)$ *can be realized by the homogeneous model* $Sp(1)AN/Sp(1)$*, where AN stands for the solvable part of the Iwasawa decomposition of the full isometry group* $Sp(n, 1)$.

In the case of $\mathbb{R}\mathrm{H}(n)$, even all the holonomy algebras of canonical connections and the types of the corresponding homogeneous structures are known, see Proposition 5 below. We first recall some definitions and notations.

Assume that $G = F_r N$ acts transitively on $\mathbb{R}\mathrm{H}(n)$ as in Proposition 2. This implies that $\mathbb{R}\mathrm{H}(n) = G/H$, with $H = F_r \cap SO(n-1)$. Then H is reductive, and thus $\mathfrak{h} = \mathfrak{h}_0 \oplus \mathfrak{h}_{ss}$, where $\mathfrak{h}_0$ is abelian and $\mathfrak{h}_{ss}$ is semisimple. Let $\mathfrak{f}_r = \mathfrak{h} \oplus \mathfrak{a}_r$, $\mathfrak{g} = \mathfrak{h} \oplus \mathfrak{a}_r \oplus \mathfrak{n}$, with $\mathfrak{a}_r$ projecting nontrivially to $\mathfrak{a} = \mathbb{R}_{>0}$. Also $\mathfrak{f}_r$ is reductive, with $\mathfrak{f}_r = (\mathfrak{h}_0 \oplus \mathfrak{a}_r) \oplus \mathfrak{h}_{ss}$. Let $\mathfrak{s} = \mathfrak{a} \oplus \mathfrak{n}$ and $\mathfrak{s}_r = \mathfrak{a}_r \oplus \mathfrak{n}$, where $\mathfrak{a}_r$ is any one-dimensional complement to $\mathfrak{h}_0 \oplus \mathfrak{n}$ in $\mathfrak{s}_f = (\mathfrak{f}_r)_0 \oplus \mathfrak{n}$. A homogeneous Riemannian structure on G/H depends on a choice of ad_H-invariant complement $\mathfrak{m}$ to $\mathfrak{h}$ in $\mathfrak{g}$, which is the graph of an $\mathfrak{h}$-equivariant map $\varphi_r\colon \mathfrak{s}_r \to \mathfrak{h}$. For any $\mathfrak{h}$-equivariant map $\chi_r\colon \mathfrak{s} \to \mathfrak{s}_r$ extending the identity on $\mathfrak{n}$, one defines $\varphi\colon \mathfrak{s} \to \mathfrak{h}$ as $\varphi = \varphi_r \circ \chi_r$. Then we have

Proposition 5 ([9, Theorems 1.1,5.2]) *The holonomy algebras of canonical connections on* $\mathbb{R}\mathrm{H}(n)$ *are* $\mathfrak{so}(n)$ *and all the reductive algebras* $\mathfrak{k} = \mathfrak{k}_0 \oplus \mathfrak{k}_{ss}$ *of compact type with* $\mathfrak{k}_0 \cong \mathbb{R}^r$ *abelian and* $\mathfrak{k}_{ss}$ *semisimple such that* $3r + \dim \mathfrak{k}_{ss} \leqslant n-1$.

Let S be a nonzero homogeneous tensor for $\mathbb{R}\mathrm{H}(n)$ *with holonomy algebra* $\mathfrak{hol}$. *Then S always has a nontrivial component in* $\mathscr{S}_1$ *and S is of type* $\mathscr{S}_1$ *if and only if* $\mathfrak{hol}$ *is* 0. *The structure is of strict type* $\mathscr{S}_{13}$ *if and only if* $\mathfrak{a} \subset \ker \varphi$ *and* $\mathfrak{hol}$ *is a nonzero semisimple algebra acting trivially on* $\ker \varphi$. *Otherwise S is of general type.*

All the homogeneous Kähler structures on the solvable description $\mathbb{C}\mathrm{H}(n) \equiv AN$ of the complex hyperbolic space have been given in a rather explicit way in [17, Theorem 3.1]. As expected, the expression simplifies a great deal for $n = 1$ and $n = 2$, which are of course interesting cases on their own.

On the other hand, the use of the parabolic subgroups of the respective full isometry groups permits us to make explicit more homogeneous descriptions and give the corresponding types of structures. In the case of $\mathbb{H}\mathrm{H}(n)$, $n = 2, 3$, one has the next result (for detailed expressions and more details see [5, Theorem 5]) and [6, Theorem 3.4]).

Proposition 6 *Let $G = KAN$ be the Iwasawa decomposition of $Sp(2, 1)$ (resp. $Sp(3, 1)$). The homogeneous descriptions of $\mathbb{H}\mathrm{H}(2)$ (resp. $\mathbb{H}\mathrm{H}(3)$) are as in the Table 1, where E is simply connected and abelian. In this case the corresponding types of homogeneous quaternionic Kähler structures are also given. The figure on the third column, if any, stands for the number of parameters of the corresponding n-parametric family of homogeneous quaternionic Kähler structures.*

Consider now the Poincaré half-space model

$$(H^n, g) = \left(\left\{ \left(u^1, \ldots, u^n\right) \in \mathbb{R}^n \,:\, u^1 > 0 \right\},\ -\frac{1}{c(u^1)^2} \sum_{i=1}^{n} \mathrm{d}u^i \otimes \mathrm{d}u^i \right)$$

of $\mathbb{R}\mathrm{H}(n)$, equipped with the metric g of constant curvature $c < 0$, and the Siegel domains

$$D_{\mathbb{C}^n} = \left\{ \left(u^1 = x + \mathrm{i}y, u^2, \ldots, u^n\right) \in \mathbb{C}^n \,:\, x - \sum_{k=2}^{n} |u^k|^2 > 0 \right\},$$

$$D_{\mathbb{H}^n} = \left\{ \left(u^1 = x + \mathrm{i}y + \mathrm{j}z + \mathrm{k}t, u^2, \ldots, u^n\right) \in \mathbb{H}^n \,:\, x - \sum_{k=2}^{n} |u^k|^2 > 0 \right\}.$$

Consider also the next vector fields on the relevant manifolds: ξ, metrically dual to the form θ in the expression of the elements of $\mathscr{S}_1$; ξ and η, metrically dual to the forms $\theta_1 + \theta_2$ and $\theta_1 - \theta_2$ in the expressions of the elements of $\mathscr{K}_2$ and $\mathscr{K}_4$; and

Table 1 Homogeneous descriptions of $\mathbb{H}H(2)$ and $\mathbb{H}H(3)$ and the corresponding types of structures

		dim E	n	Type
$Sp(2,1)/(Sp(2) \times Sp(1))$		0		$\{0\}$
$E_{\lambda,\mu}N$	$(\lambda, \mu \in \mathbb{R}^3 \setminus \{0\})$	1	6	$\mathscr{QK}_{12345}$
$E_{0,\mu}N$	$(\mu \in \mathbb{R}^3 \setminus \{0\})$	1	3	$\mathscr{QK}_{1345}$
$AN = E_{0,0}N$		1		$\mathscr{QK}_{134}$
$Sp(3,1)/(Sp(3) \times Sp(1))$		0		$\{0\}$
$E_{\lambda,\mu,\nu,\gamma}N$	$(\lambda, \mu, \nu \in \mathbb{R}^3 \setminus \{0\},\ \gamma \in \mathbb{R}^4 \setminus \{0\})$	1	13	$\mathscr{QK}_{12345}$
$E_{0,\mu,\nu,\gamma}N$	$(\mu, \nu \in \mathbb{R}^3 \setminus \{0\},\ \gamma \in \mathbb{R}^4 \setminus \{0\})$	1	10	$\mathscr{QK}_{1345}$
$AN = E_{0,0,0,0}N$		1		$\mathscr{QK}_{134}$

ξ, metrically dual to the form θ in the expression of the elements of $\mathscr{Q}\mathscr{K}_3$. Each of these vector fields ξ defines the corresponding homogeneous structures of linear type and we have the next result.

Proposition 7 ([25, (5.26)], [16, (4.4)], [4, Corollary 5.1]) *Consider the Poincaré half-space model (H^n, g) of $\mathbb{R}\mathrm{H}(n)$ (resp. Siegel domain model $(D_{\mathbb{C}^n}, g)$ or $(D_{\mathbb{H}^n}, g)$ of $\mathbb{C}\mathrm{H}(n)$ or $\mathbb{H}\mathrm{H}(n)$), with $x = \operatorname{Re} u^1$ the respective distinguished real coordinate. Then each of the vector fields,*

$$\xi = -cx\frac{\partial}{\partial x}$$

on (H^n, g) (see Fig. 1*) and*

$$\xi = -\frac{c}{2}\left(x - \sum_{k=2}^{n} |u^k|^2\right)\frac{\partial}{\partial x}, \qquad c < 0,$$

on $(D_{\mathbb{C}^n}, g)$ or $(D_{\mathbb{H}^n}, g)$ (see Fig. 2*), defines a homogeneous structure of linear type, c being the ordinary, holomorphic or quaternionic sectional curvature, respectively.*

The expressions of the vector fields ξ for the open unit ball models of $\mathbb{R}\mathrm{H}(n)$, $\mathbb{C}\mathrm{H}(n)$ and $\mathbb{H}\mathrm{H}(n)$ are rather more complicated than those for the previous models, as one may see in the next proposition.

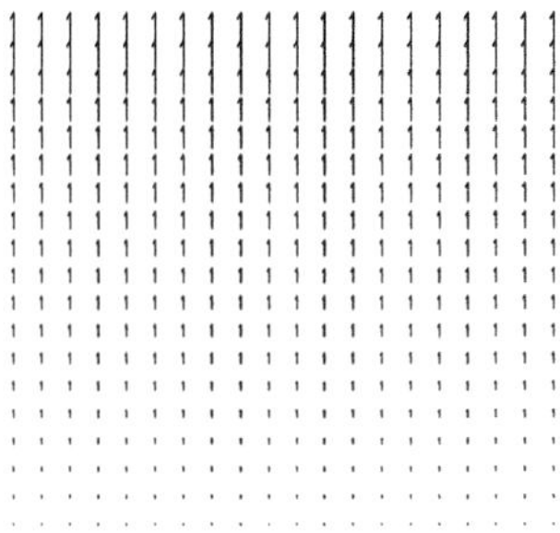

Fig. 1 The vector field ξ on the Poincaré half-space model of $\mathbb{R}\mathrm{H}(n)$

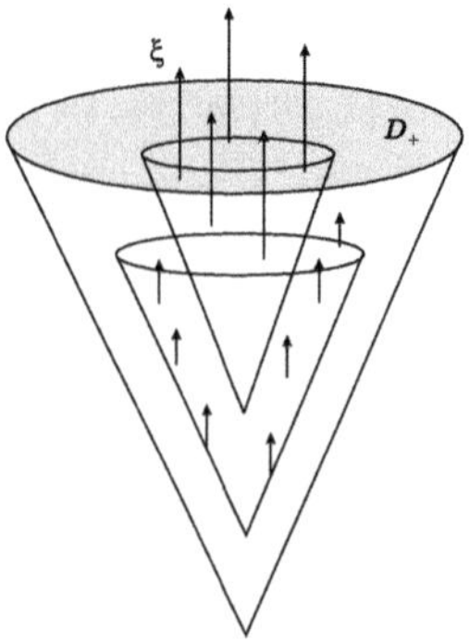

Fig. 2 The vector field ξ on the Siegel domain model of $\mathbb{C}\mathrm{H}(n)$ or $\mathbb{H}\mathrm{H}(n)$

Proposition 8 **(i)** ([4, Proposition 2.7]) *The open unit ball model* (B^n, g) *of* $\mathbb{R}\mathrm{H}(n)$ *with negative constant sectional curvature c, admits a homogeneous Riemannian structure of linear type defined by the vector field*

$$\xi_{B^n} = \frac{c\left(1-\sum_{i=1}^n (x^i)^2\right)}{2\left((1+x^1)^2+\sum_{i=2}^n (x^i)^2\right)}\left(\left((1+x^1)^2-\sum_{i=2}^n (x^i)^2\right)\frac{\partial}{\partial x^1}+2(1+x^1)\sum_{i=2}^n x^i\frac{\partial}{\partial x^i}\right).$$

(ii) ([4, Proposition 3.11]) *The open unit ball model* (B^n, g, J) *of* $\mathbb{C}\mathrm{H}(n)$ *with negative constant holomorphic sectional curvature c, admits a homogeneous Kähler structure of linear type defined by the vector field*

$$\begin{aligned}\xi_{B^n} = -\frac{c\left(1-\sum_k ((x^k)^2)+(y^k)^2\right)}{4\left((1-x^1)^2+(y^1)^2\right)}&\left(\left((1-x^1)^2-(y^1)^2\right)\frac{\partial}{\partial x^1}+2(x^1-1)y^1\frac{\partial}{\partial y^1}\right.\\ &+\sum_{k=2}^n\left(\left((x^1-1)x^k-y^1y^k\right)\frac{\partial}{\partial x^k}\right.\\ &\left.\left.+\left((x^1-1)y^k+y^1x^k\right)\frac{\partial}{\partial y^k}\right)\right).\end{aligned}$$

(iii) ([4, Proposition 4.11]) *The open unit ball model* (B^n, g, υ^3) *of* $\mathbb{H}\mathrm{H}(n)$ *with negative constant quaternionic sectional curvature c admits a homogeneous quaternionic Kähler structure S of linear type defined by the vector field*

$$\begin{aligned}\xi_{B^n} = &-\frac{c\left(1-\sum_{k=1}^n |q^k|^2\right)}{4|q^1-1|^2}\\ &\times\left\{((x^1-1)^2-(y^1)^2-(z^1)^2-(w^1)^2)\frac{\partial}{\partial x^1}+2(x^1-1)\left(y^1\frac{\partial}{\partial y^1}+z^1\frac{\partial}{\partial z^1}+w^1\frac{\partial}{\partial w^1}\right)\right.\\ &-\sum_{k=2}^n\left((1-x^1)x^k+y^1y^k+z^1z^k+w^1w^k)\frac{\partial}{\partial x^k}+((1-x^1)y^k-y^1x^k+z^1w^k-w^1z^k)\frac{\partial}{\partial y^k}\right.\\ &\left.\left.+((1-x^1)z^k-y^1w^k-z^1x^k+w^1y^k)\frac{\partial}{\partial z^k}+((1-x^1)w^k+y^1z^k-z^1y^k-w^1x^k)\frac{\partial}{\partial w^k}\right)\right\}.\end{aligned}$$

Remark 1 We finally point out two open problems:

(a) The analogues of Proposition 5 for $\mathbb{C}\mathrm{H}(n)$ and $\mathbb{H}\mathrm{H}(n)$.

(b) The characterization of the Cayley hyperbolic plane $\mathbb{O}\mathrm{H}(2)$ by suitable defined homogeneous Spin(9)-structures (see Friedrich [13, 14], Mykytyuk [20, 21]).

Acknowledgments The second author has been supported by the Ministry of Economy and Competitiveness, Spain, under Project MTM2013-46961-P.

References

1. Abbena, E., Garbiero, S.: Almost Hermitian homogeneous structures. P. Edinburgh Math. Soc. **31**(3), 375–395 (1988)
2. Alekseevsky, D.V., Cortés, V.: Isometry groups of homogeneous quaternionic Kähler manifolds. J. Geom. Anal. **9**(4), 513–545 (1999)
3. Ambrose, W., Singer, I.M.: On homogeneous Riemannian manifolds. Duke Math. J. **25**, 647–669 (1958)
4. Batat, W., Gadea, P.M., Oubiña, J.A.: Homogeneous pseudo-Riemannian structures of linear type. J. Geom. Phys. **61**(3), 745–764 (2011)
5. Castrillón López, M., Gadea, P.M., Oubiña, J.A.: Homogeneous quaternionic Kähler structures on 12-dimensional Alekseevsky spaces. J. Geom. Phys. **57**(10), 2098–2113 (2007)
6. Castrillón López, M., Gadea, P.M., Oubiña, J.A.: Homogeneous quaternionic Kähler structures on eight-dimensional non-compact quaternion-Kähler symmetric spaces. Math. Phys. Anal. Geom. **12**(1), 47–74 (2009)
7. Castrillón López, M., Gadea, P.M., Swann, A.F.: Homogeneous quaternionic Kähler structures and quaternionic hyperbolic space. Transform. Groups **11**(4), 575–608 (2006)
8. Castrillón López, M., Gadea, P.M., Swann, A.F.: Homogeneous structures on the real and complex hyperbolic spaces. Illinois J. Math. **53**, 561–574 (2009)
9. Castrillón López, M., Gadea, P.M., Swann, A.F.: The homogeneous geometries of real hyperbolic space. Mediterr. J. Math. **10**(2), 1011–1022 (2013)
10. Chen, S.S., Greenberg, L.: Hyperbolic spaces. In: Contributions to Analysis (a collection of papers dedicated to Lipman Bers), pp. 49–87. Academic Press, New York (1974)
11. Fino, A.: Intrinsic torsion and weak holonomy. Math. J. Toyama Univ. **21**, 1–22 (1998)
12. Fortuny, P., Gadea, P.M.: On the classification theorem of almost-Hermitian or homogeneous Kähler structures. Rocky Mt. J. Math. **36**, 213–223 (2006)
13. Friedrich, Th.: Weak Spin(9)-structures on16-dimensional Riemannian manifolds. Asian J. Math. **5**(1), 129–160 (2001)
14. Friedrich, Th.: Spin(9)-structures and connections with totally skew-symmetric torsion. J. Geom. Phys. **47**(2–3), 197–206 (2003)
15. Gadea, P.M., González-Dávila, J.C., Oubiña, J.A.: Homogeneous Riemannian manifolds with the simplest Dirac operator, preprint (2015)
16. Gadea, P.M., Montesinos Amilibia, A., Muñoz Masqué, J.: Characterizing the complex hyperbolic space by Kähler homogeneous structures. Math. Proc. Camb. **128**(1), 87–94 (2000)
17. Gadea, P.M., Oubiña, J.A.: Homogeneous Kähler and Sasakian structures related to complex hyperbolic spaces. P. Edinb. Math. Soc. **53**(2), 393–413 (2010)
18. Heintze, E.: On homogeneous manifolds of negative curvature. Math. Ann. **211**, 23–34 (1974)
19. Ishihara, S.: Quaternion Kählerian manifolds. J. Differ. Geom. **9**, 483–500 (1974)
20. Mykytyuk, I.V.: Kähler structures on tangent bundles of symmetric spaces of rank one. Sb. Math. **192**(11–12), 1677–1704 (2001)
21. Mykytyuk, I.V.: The triple Lie system of the symmetric space $F_4/\mathrm{Spin}(9)$. Asian J. Math. **6**(4), 713–718 (2002)
22. Ratcliffe, J.G.: Foundations of Hyperbolic Manifolds, 2nd edn. Springer, New York (2006)
23. Salamon, S.: Riemannian Geometry and Holonomy Groups. Longman Sci. & Tech, Harlow (1989)
24. Sekigawa, K.: Notes on homogeneous almost Hermitian manifolds. Hokkaido Math. J. **7**, 206–213 (1978)
25. Tricerri, F., Vanhecke, L.: Homogeneous Structures on Riemannian Manifolds. Lond. Math. Soc. Lect. Note Ser. 83. Cambridge Univ. Press, Cambridge, (1983)
26. Weyl, H.: The Classical Groups. Princeton University Press, Princeton (1939)

On the (1 + 3) Threading of Spacetime

Aurel Bejancu

Dedicated to Jaime Muñoz Masqué on the occasion of his 65th birthday

Abstract We develop a $(1 + 3)$ threading formalism of the spacetime with respect to a non-normalized timelike vector field. It is worth mentioning that in our approach the spatial distribution is not necessarily integrable. Thus, this formalism is suitable for general Lorentz metrics from both the theory of black holes and perturbation theory. Also, the simple form of the $(1 + 3)$ decomposition of Einstein Field Equations stated in the paper, might have an important impact on the work of discovering new inhomogeneous cosmological models.

Keywords (1+3) Threading formalism · (1 +3) Decomposition of Einstein field equations · Riemannian spatial connection · Spatial tensor fields

1 Introduction

In cosmology, in order to relate the physics and geometry to the observations, it is frequently used the $(1 + 3)$ threading of spacetime. Namely, it is taken a unit 4-velocity field **u** which is tangent to a preferred congruence of world lines. Then, the study of both physics and geometry of the spacetime is developed by considering (provided they exist), orthogonal hypersurfaces to **u**. This was successfully applied to the study of the Friedmann–Lemaître–Robertson–Walker universe (cf. [4]). Also, the gravito-electromagnetism and the splitting of Einstein Field Equations (EFE)

A. Bejancu (✉)
Department of Mathematics, Kuwait University, P.O. Box 5969, 13060 Safat, Kuwait
e-mail: aurel.bejancu@ku.edu.kw

M. Castrillón López et al. (eds.), *Geometry, Algebra and Applications: From Mechanics to Cryptography*, Springer Proceedings in Mathematics & Statistics 161, DOI 10.1007/978-3-319-32085-4_2

have been intensively studied (cf. [6, 8–10]). Two conditions have been imposed on the geometric objects of the spacetime:

(i) **u** must be a unit vector field.

(ii) The distribution that is orthogonal to the congruence determined by **u**, must be integrable.

Recently, we developed a new point of view on the $(1 + 3)$ threading of spacetime, where we removed both conditions above (cf. [1, 2]). In this general setting we obtained in a covariant form, the fully general $3D$ equations of motion and a $3D$ identity satisfied by the geodesics of a spacetime. Also, we applied this general method to the study of Kerr-Newman black holes.

The main purpose of this paper is to state the $(1 + 3)$ decomposition of EFE with respect to arbitrary timelike vector field and spatial distribution. The study is based on both the Riemannian spatial connection and the spatial tensor fields defined in [2]. It is worth mentioning that each group of the EFE given by (7.3) is invariant with respect to the transformations of coordinates on the spacetime.

Now, we outline the content of the paper. In Sect. 2 we introduce the kinematic quantities determined by a non-normalized timelike vector field ξ. Note that in [2] we put on Φ given by (2.2a) the condition that it is independent of time. This condition is satisfied by all stationary black holes (cf. [3, 5]). However, in perturbation theory (cf. [7, 11]), Φ is not, in general, independent of time. For the sake of general applications of our study, we remove the above condition on Φ. In Sect. 3 we present the Riemannian spatial connection and express the Levi-Civita connection of the spacetime (M, g) in terms of spatial tensor fields (cf. (3.4)). The local coefficients of the Riemannian spatial connection and the kinematic quantities are used in Sect. 4 to express the fully general equations of motion in (M, g). In particular, we obtain a geometric characterization of the spatial geodesics. In Sect. 5 we obtain the structure equations for the spatial distribution (cf. (5.1)), which lead us to the decomposition (6.6) of the Ricci tensor of (M, g). Also, in Sect. 6 we deduce the Raychaudhuri equation (6.8) for the $(1 + 3)$ threading formalism determined by ξ, and find the local components of the stress-energy-momentum tensor field with respect to the threading frame field (cf. (6.12)). Finally, we state the $(1 + 3)$ decomposition of the EFE (cf. (7.3)).

2 Kinematic Quantities in a Spacetime with Respect to a Non-Normalized Timelike Vector Field

Let (M, g) be a $4D$ spacetime, and ξ be a timelike vector field on M which is not necessarily normalized. Then, we have

$$TM = VM \oplus SM, \tag{2.1}$$

where VM is the *time distribution* spanned by ξ and SM is the *spatial distribution* that is complementary orthogonal to VM in TM.

Throughout the paper we use the ranges of indices: $i, j, k, \dots \in \{1, 2, 3\}$ and $a, b, c, \ldots \in \{0, 1, 2, 3\}$. Also, for any vector bundle E over M, denote by $\Gamma(E)$ the $\mathscr{F}(M)$-module of smooth sections of E, where $\mathscr{F}(M)$ is the algebra of smooth functions on M.

Now, we consider a coordinate system (x^a) on M such that $\xi = \partial/\partial x^0$. Then, we put

$$(a)\quad \xi_0 = g\left(\frac{\partial}{\partial x^0}, \frac{\partial}{\partial x^0}\right) = -\Phi^2, \quad (b)\quad \xi_i = g\left(\frac{\partial}{\partial x^i}, \frac{\partial}{\partial x^0}\right), \qquad (c)\quad g_{ij} = g\left(\frac{\partial}{\partial x^i}, \frac{\partial}{\partial x^j}\right), \tag{2.2}$$

where Φ is a non-zero function that is globally defined on M. The decomposition (2.1) enables us to use the *threading frame* $\{\partial/\partial x^0, \delta/\delta x^i\}$ and the *threading coframe field* $\{\delta x^0, dx^i\}$, where we put (cf. [1, 2])

$$(a)\quad \frac{\delta}{\delta x^i} = \frac{\partial}{\partial x^i} + \Phi^{-2}\xi_i\frac{\partial}{\partial x^0}, \quad (b)\quad \delta x^0 = dx^0 - \Phi^{-2}\xi_i dx^i. \tag{2.3}$$

The Lie brackets of the vector fields from the threading frame are expressed as follows:

$$(a)\quad \left[\frac{\delta}{\delta x^j}, \frac{\delta}{\delta x^i}\right] = 2\omega_{ij}\frac{\partial}{\partial x^0}, \quad (b)\quad \left[\frac{\partial}{\partial x^0}, \frac{\delta}{\delta x^i}\right] = a_i\frac{\partial}{\partial x^0}, \tag{2.4}$$

where we put

$$(a)\quad \omega_{ij} = \Phi^{-2}\left\{c_i\xi_j - c_j\xi_i + \frac{1}{2}\left(\frac{\delta\xi_i}{\delta x^j} - \frac{\delta\xi_j}{\delta x^i}\right)\right\},$$
$$(b)\quad c_i = \Phi^{-1}\frac{\delta\Phi}{\delta x^i}, \qquad (c)\quad a_i = \Phi^{-2}\left\{\frac{\partial\xi_i}{\partial x^0} - 2\Psi\xi_0\right\}, \tag{2.5}$$
$$(d)\quad \Psi = \Phi^{-1}\frac{\partial\Phi}{\partial x^0}.$$

As SM is integrable if and only if $\omega_{ij} = 0$ for all $i, j \in \{1, 2, 3\}$, we say that $\{\omega_{ij}\}$ are the local components of the *vorticity tensor field* with respect to the threading frame.

Next, we denote by h_{ij} the local components of the Riemannian metric induced by g on SM with respect to the threading frame, and deduce that

$$h_{ij} = g\left(\frac{\delta}{\delta x^i}, \frac{\delta}{\delta x^j}\right) = g_{ij} + \Phi^{-2}\xi_i\xi_j. \tag{2.6}$$

Thus the line element of the Lorentz metric g on M with respect to the threading coframe is expressed as follows:

$$ds^2 = -\Phi^2(\delta x^0)^2 + h_{ij}dx^i dx^j. \tag{2.7}$$

By using h_{ij} and the entries h^{ij} of the inverse of the 3×3 matrix $[h_{ij}]$, we define the expansion tensor field $\{\Theta_{ij}\}$, the expansion function Θ, and the shear tensor field $\{\sigma_{ij}\}$, as follows

$$(a)\ \Theta_{ij} = \frac{1}{2}\frac{\partial h_{ij}}{\partial x^0}, \quad (b)\ \Theta = h^{ij}\Theta_{ij}, \quad (c)\ \sigma_{ij} = \Theta_{ij} - \frac{1}{3}\Theta h_{ij}. \tag{2.8}$$

Raising and lowering indices $i, j, k, \ldots$ are done by using h^{ij} and h_{ij}, as for example:

$$(a)\ \omega_j^k = h^{ki}\omega_{ij}, \quad (b)\ \omega_{ij} = h_{ik}\omega_j^k, \quad (c)\ \omega^{kl} = h^{ki}h^{lj}\omega_{ij}. \tag{2.9}$$

In earlier literature, spatial tensor fields have been introduced as projections on SM of the tensor fields on M (cf. [4, 6, 8]). In our approach, *a spatial tensor field* of type (p, q) is locally given by 3^{p+q} locally defined functions $T_{i\cdots}^{k\cdots}$, satisfying

$$T_{i\cdots}^{k\cdots}\frac{\partial \tilde{x}^h}{\partial x^k}\cdots = \tilde{T}_{j\cdots}^{h\cdots}\frac{\partial \tilde{x}^j}{\partial x^i}\cdots,$$

with respect to the coordinate transformations $\tilde{x}^a = \tilde{x}^a(x^0, x^i)$ on M. It is worth mentioning that $\{h_{ij}, \omega_{ij}, \Theta_{ij}, \sigma_{ij}\}$ and $\{a_i, c_i\}$ are spatial tensor fields of type $(0, 2)$ and $(0, 1)$, respectively.

3 The Riemannian Spatial Connection on a Spacetime

Let ∇ be the Levi-Civita connection on the spacetime (M, g). Then the *Riemannian spatial connection* on M is a metric linear connection $\nabla^\star$ on SM, given by

$$(a)\ \nabla_X^\star sY = s\nabla_X sY, \quad \forall\ X, Y \in \Gamma(SM), \tag{3.1}$$

where s is the projection morphism of TM on SM. Locally, $\nabla^\star$ is given by

$$(a)\ \nabla^\star_{\frac{\delta}{\delta x^j}}\frac{\delta}{\delta x^i} = \Gamma_{i\ j}^{\star\ k}\frac{\delta}{\delta x^k}, \quad (b)\ \nabla^\star_{\frac{\partial}{\partial x^0}}\frac{\delta}{\delta x^i} = \Gamma_{i\ 0}^{\star\ k}\frac{\delta}{\delta x^k}, \tag{3.2}$$

where we put

$$
\begin{aligned}
&(a)\quad \Gamma^{\star}{}_{i}{}^{k}{}_{j} = \frac{1}{2}h^{kl}\left\{\frac{\delta h_{lj}}{\delta x^i} + \frac{\delta h_{li}}{\delta x^j} - \frac{\delta h_{ij}}{\delta x^l}\right\},\\
&(b)\quad \Gamma^{\star}{}_{i}{}^{k}{}_{0} = \Theta_i^k + \Phi^2\omega_i^k.
\end{aligned}
\tag{3.3}
$$

Remark 3.1 The Riemannian spatial connection $\nabla^{\star}$ is different from the *three-dimensional operator* $\bar{\nabla}$ that has been used in earlier literature (cf. (4.19) of [4]). Note that $\nabla^{\star}$ is a metric linear connection on SM, and therefore defines covariant derivatives of spatial tensor fields with respect to vector fields on M. On the contrary, $\bar{\nabla}$ is an operator which acts on tensor fields on M, but it does not define a linear connection on M. □

Next, by using (3.1)–(3.3), we express the Levi-Civita connection on (M, g), as follows:

$$
\begin{aligned}
&(a)\quad \nabla_{\frac{\delta}{\delta x^j}}\frac{\delta}{\delta x^i} = \Gamma^{\star}{}_{i}{}^{k}{}_{j}\frac{\delta}{\delta x^k} + \left(\omega_{ij} + \Phi^{-2}\Theta_{ij}\right)\frac{\partial}{\partial x^0},\\
&(b)\quad \nabla_{\frac{\partial}{\partial x^0}}\frac{\delta}{\delta x^i} = \left(\Theta_i^k + \Phi^2\omega_i^k\right)\frac{\delta}{\delta x^k} + b_i\frac{\partial}{\partial x^0},\\
&(c)\quad \nabla_{\frac{\delta}{\delta x^i}}\frac{\partial}{\partial x^0} = \left(\Theta_i^k + \Phi^2\omega_i^k\right)\frac{\delta}{\delta x^k} + c_i\frac{\partial}{\partial x^0},\\
&(d)\quad \nabla_{\frac{\partial}{\partial x^0}}\frac{\partial}{\partial x^0} = \Phi^2 b^k\frac{\delta}{\delta x^k} + \Psi\frac{\partial}{\partial x^0},
\end{aligned}
\tag{3.4}
$$

where we put

$$
b_i = a_i + c_i. \tag{3.5}
$$

Denote by $R^{\star}$ the curvature tensor field of $\nabla^{\star}$ and put

$$
\begin{aligned}
&(a)\quad R^{\star}\left(\frac{\delta}{\delta x^h}, \frac{\delta}{\delta x^k}, \frac{\delta}{\delta x^i}\right) = R^{\star j}_{i\ kh}\frac{\delta}{\delta x^j},\\
&(b)\quad R\left(\frac{\delta}{\delta x^0}, \frac{\delta}{\delta x^k}, \frac{\delta}{\delta x^i}\right) = R^{\star j}_{i\ k0}\frac{\delta}{\delta x^j}.
\end{aligned}
$$

Then by using (3.1)–(3.3), (2.4) and the well-known formula for $R^{\star}$, we obtain

$$
\begin{aligned}
&(a)\quad R^{\star j}_{i\ kh} = \frac{\delta \Gamma^{\star}{}_{i}{}^{j}{}_{k}}{\delta x^h} - \frac{\delta \Gamma^{\star}{}_{i}{}^{j}{}_{h}}{\delta x^k} + \Gamma^{\star}{}_{i}{}^{l}{}_{k}\Gamma^{\star}{}_{l}{}^{j}{}_{h} - \Gamma^{\star}{}_{i}{}^{l}{}_{h}\Gamma^{\star}{}_{l}{}^{j}{}_{k} - 2\omega_{kh}\Gamma^{\star}{}_{i}{}^{j}{}_{0},\\
&(b)\quad R^{\star j}_{i\ k0} = \frac{\partial \Gamma^{\star}{}_{i}{}^{j}{}_{k}}{\partial x^0} - \Gamma^{\star}{}_{i}{}^{j}{}_{0|k} - a_k\Gamma^{\star}{}_{i}{}^{j}{}_{0}.
\end{aligned}
\tag{3.6}
$$

Here, and in the sequel, the vertical bar "|" represents covariant derivative with respect to the Riemannian spatial connection.

4 3*D* Equations of Motion in a 4*D* Spacetime

Let C be a smooth curve in M given by parametric equations

$$x^a = x^a(t), \quad a \in \{0, 1, 2, 3\}, \quad t \in [\alpha, \beta],$$

where (x^a) is the special coordinate system introduced by the $(1+3)$ threading of (M, g). The tangent vector field d/dt to C is expressed as follows:

$$\frac{d}{dt} = \frac{dx^i}{dt}\frac{\delta}{\delta x^i} + \frac{\delta x^0}{\delta t}\frac{\partial}{\partial x^0}, \tag{4.1}$$

where we put

$$\frac{\delta x^0}{\delta t} = \frac{dx^0}{dt} - \Phi^{-2}\xi_i\frac{dx^i}{dt}.$$

By direct calculations, using (4.1) and (3.4) we deduce that C is a geodesic of (M, g), if and only if,

$$\begin{aligned} &(a)\quad \frac{d^2x^k}{dt^2} + \Gamma^{\star}{}_{i}{}^{k}{}_{j}\frac{dx^i}{dt}\frac{dx^j}{dt} + 2\frac{\delta x^0}{\delta t}\frac{dx^i}{dt}(\Theta_i^k + \Phi^2\omega_i^k) + \Phi^2\left(\frac{\delta x^0}{\delta t}\right)^2 b^k = 0, \\ &(b)\quad \frac{d}{dt}\left(\frac{\delta x^0}{\delta t}\right) + \Phi^{-2}\Theta_{ij}\frac{dx^i}{dt}\frac{dx^j}{dt} + \frac{\delta x^0}{dt}(b_i + c_i)\frac{dx^i}{dt} + \left(\frac{\delta x^0}{dt}\right)^2\Psi = 0. \end{aligned} \tag{4.2}$$

We note that Eq. (4.2) represent the splitting of the fully general equations of motion of the spacetime. We call (4.2a) the 3*D* *equations of motion* in the 4*D* spacetime (M, g). It is worth mentioning that these equations are related to the equations of autoparallel curves of the Riemannian spatial connection. To show this we introduce a special class of geodesics in (M, g). A geodesic C of (M, g) is called a *spatial geodesic*, if it satisfies one of the following conditions:

$$(a)\quad \frac{\delta x^0}{\delta t} = 0 \quad \text{or} \quad (b)\quad \frac{d}{dt} = \frac{dx^i}{dt}\frac{\delta}{\delta x^i}. \tag{4.3}$$

Taking into account (4.2) and (4.3), we deduce that a curve C is a spatial geodesic, if and only if, (4.3) and the following equations are satisfied:

$$\begin{aligned} &(a)\quad \frac{d^2x^k}{dt^2} + \Gamma^{\star k}_{i\ j}\frac{dx^i}{dt}\frac{dx^j}{dt} = 0, \\ &(b)\quad \Theta_{ij}\frac{dx^i}{dt}\frac{dx^j}{dt} = 0. \end{aligned} \tag{4.4}$$

Now, we say that a curve C in M is autoparallel for the Riemannian spatial connection $\nabla^{\star}$, if it satisfies (4.3) and

$$\nabla^{\star}_{\frac{d}{dt}} \frac{d}{dt} = 0. \tag{4.5}$$

By using (4.3b) and (3.2a) into (4.5) we infer that C is an autoparallel for $\nabla^{\star}$, if and only if, (4.3a) and (4.4a) are satisfied. Now, from (3.4a) we see that

$$K_{ij} = \omega_{ij} + \Phi^{-2}\Theta_{ij}, \tag{4.6}$$

can be thought as local components of the second fundamental form of SM. Then we say that a curve C in M is an asymptotic line for SM if it satisfies (4.3) and the following equation

$$K_{ij}\frac{dx^i}{dt}\frac{dx^j}{dt} = 0. \tag{4.7}$$

Taking into account that ω_{ij} define a skew-symmetric spatial tensor field, and using (4.6) into (4.7), we deduce that C is an asymptotic line for SM, if and only if, it satisfies (4.3) and (4.4b). Summing up these results, we can state the following:

A curve C in a spacetime (M, g) is a spatial geodesic, if and only if, the following conditions are satisfied:

(i) *C is autoparallel for the Riemannian spatial connection.*
(ii) *C is an asymptotic line for the spatial distribution.*

5 Structure Equations for the Spatial Distribution

In this section, R stands for both curvature tensor fields of types (0,4) and (1,3) of ∇, related by

$$R(X, Y, Z, U) = g(R(X, Y, U), Z).$$

Locally, R is completely determined by the following local components with respect to the threading frame field $\{\partial/\partial x^0, \delta/\delta x^i\}$:

$$R_{ijkh} = R\Big(\frac{\delta}{\delta x^h}, \frac{\delta}{\delta x^k}, \frac{\delta}{\delta x^j}, \frac{\delta}{\delta x^i}\Big),$$

$$R_{i0kh} = R\Big(\frac{\delta}{\delta x^h}, \frac{\delta}{\delta x^k}, \frac{\partial}{\partial x^0}, \frac{\delta}{\delta x^i}\Big),$$

$$R_{i0k0} = R\Big(\frac{\partial}{\partial x^0}, \frac{\delta}{\delta x^k}, \frac{\partial}{\partial x^0}, \frac{\delta}{\delta x^i}\Big).$$

Then, by direct calculations, using (3.4) and (2.4), we deduce that

$$\begin{aligned}
(a)\quad R_{ijkh} &= R^{\star}_{ijkh} + (\Theta_{ik} + \Phi^2\omega_{ik})(\omega_{jh} + \Phi^{-2}\Theta_{jh}) \\
&\quad - (\Theta_{ih} + \Phi^2\omega_{ih})(\omega_{jk} + \Phi^{-2}\Theta_{jk}), \\
(b)\quad R_{i0kh} &= \Theta_{ih|_k} - \Theta_{ik|_h} + \Theta_{ik}c_h - \Theta_{ih}c_k \\
&\quad + \Phi^2\left\{\omega_{ih|_k} - \omega_{ik|_h} + \omega_{ih}c_k - \omega_{ik}c_h + 2\omega_{kh}b_i\right\}, \\
(c)\quad R_{i0k0} &= \Phi^2\Big\{b_ib_k + \frac{1}{2}(b_{i|k} + b_{k|i})\Big\} + \Psi\Theta_{ik} - \Theta_{ik|0} \\
&\quad - \Theta_{il}\Theta^l_k - \Phi^4\omega_{il}\omega^l_k.
\end{aligned} \tag{5.1}$$

We call (5.1) the *structure equations* for the spatial distribution *SM* on the spacetime (M, g). Note that these equations are obtained in the most general spacetime, that is, *SM* is not necessarily an integrable distribution and ξ is not necessarily a unit vector field. Such general spacetimes are intensively studied in perturbation theory (cf. [7, 11]), and the theory of black holes [3, 5].

In particular, suppose that ξ is a unit vector field, that is, we have $\Phi^2 = 1$. Then by using (2.5) and (3.5), we deduce that

$$\Psi = 0, \quad c_i = 0, \quad b_i = a_i = \frac{\partial \xi_i}{\partial x^0}, \quad \forall\, i \in \{1, 2, 3\}. \tag{5.2}$$

Hence, in this particular case, the above structure equations become

$$\begin{aligned}
(a)\quad R_{ijkh} &= R^{\star}_{ijkh} + (\Theta_{ik} + \omega_{ik})(\Theta_{jh} + \omega_{jh}) \\
&\quad - (\Theta_{ih} + \omega_{ih})(\Theta_{jk} + \omega_{jk}), \\
(b)\quad R_{i0kh} &= \omega_{ih|_k} - \omega_{ik|_h} + \Theta_{ih|_k} - \Theta_{ik|_h} + 2\omega_{kh}a_i, \\
(c)\quad R_{i0k0} &= a_ia_k + \frac{1}{2}(a_{i|_k} + a_{k|_i}) - \Theta_{ik|_0} - \Theta_{il}\Theta^l_k - \omega_{il}\omega^l_k.
\end{aligned} \tag{5.3}$$

If moreover, *SM* is an integrable distribution, that is, the vorticity tensor field vanishes identically on M, then (5.3) becomes

$$\begin{aligned}
(a)\quad R_{ijkh} &= R^{\star}_{ijkh} + \Theta_{ik}\Theta_{jh} - \Theta_{ih}\Theta_{jk}, \\
(b)\quad R_{i0kh} &= \Theta_{ih|_k} - \Theta_{ik|_h}, \\
(c)\quad R_{i0k0} &= a_ia_k + \frac{1}{2}\left(a_{i|_k} + a_{k|_i}\right) - \Theta_{ik|_0} - \Theta_{il}\Theta^l_k.
\end{aligned} \tag{5.4}$$

Finally, note that (5.4) refers to a spacetime more general than the Friedmann–Lemaître–Robertson–Walker (FLRW) universe, where we have $a_i = 0$ and $\theta_{ij} = a(x^0)h_{ij}$.

6 Ricci Tensor and Stress-Energy-Momentum Tensor

The purpose of this section is to express both the Ricci tensor of (M, g) and the stress-energy-momentum tensor in terms of spatial tensor fields. First, we consider an orthonormal frame field $\{E_k\}$ in $\Gamma(SM)$:

$$E_k = E_k^i \frac{\delta}{\delta x^i}, \tag{6.1}$$

and obtain

$$h^{ij} = \sum_{k=1}^{3} E_k^i E_k^j. \tag{6.2}$$

The Ricci tensor *Ric* of (M, g) is given by (cf. [12], p. 87)

$$Ric(X, Y) = \sum_{k=1}^{3} R(E_k, X, E_k, Y) - \Phi^{-2} R\Big(\frac{\partial}{\partial x^0}, X, \frac{\partial}{\partial x^0}, Y\Big), \tag{6.3}$$

for all $X,\ Y \in \Gamma(TM)$. Then, by using (6.1)–(6.3), we obtain

$$\begin{aligned} &(a)\ \ R_{ik} = h^{jl} R_{ijkl} - \Phi^{-2} R_{i0k0}, \quad (b)\ \ R_{i0} = h^{jl} R_{j0li}, \\ &\qquad\qquad (c)\ \ R_{00} = h^{jl} R_{j0l0}, \end{aligned} \tag{6.4}$$

where we put

$$\begin{aligned} &(a)\ \ R_{ik} = Ric\left(\frac{\delta}{\delta x^i}, \frac{\delta}{\delta x^k}\right), \quad (b)\ \ R_{i0} = Ric\left(\frac{\delta}{\delta x^i}, \frac{\partial}{\partial x^0}\right), \\ &(c)\ \ R_{00} = Ric\left(\frac{\partial}{\partial x^0}, \frac{\partial}{\partial x^0}\right). \end{aligned} \tag{6.5}$$

Next, by direct calculations using (5.1) and (6.4), we deduce that the Ricci tensor of (M, g) is given by

$$\begin{aligned} (a)\ \ R_{ik} &= R^{\star}_{ik} - b_i b_k - \frac{1}{2}\left(b_{i|_k} + b_{k|_i}\right) \\ &\quad + \Phi^{-2}\left\{\Theta_{ik|_0} + (\Theta - \Psi)\Theta_{ik}\right\}, \\ (b)\ \ R_{i0} &= \Theta^j_{i|_j} - \Theta_{|_i} + \Theta c_i - \Theta^j_i c_j \\ &\quad + \Phi^2\left\{\omega^j_{i\ |_j} + \omega^j_i c_j + 2\omega^j_i b_j\right\}, \\ (c)\ \ R_{00} &= -\Theta_{|_0} - \Theta_{ij}\Theta^{ij} + \Psi\Theta + \Phi^2\left\{b_j b^j + b^j_{\ |_j} + \Phi^2 \omega^2\right\}, \end{aligned} \tag{6.6}$$

where we put

$$R^{\star}_{ik} = \frac{1}{2}\left(R^{\star l}_{i\ kl} + R^{\star l}_{k\ il}\right), \quad \Theta_{|i} = \frac{\delta \Theta}{\delta x^i}, \quad \Theta_{|0} = \frac{\partial \Theta}{\partial x^0}, \quad \omega^2 = \omega_{ij}\omega^{ij}.$$

Taking into account (2.8b) and (2.8c), we deduce that

$$\Theta_{ij}\Theta^{ij} = \sigma^2 + \frac{1}{3}\Theta^2, \tag{6.7}$$

where we put

$$\sigma^2 = \sigma_{ij}\sigma^{ij}.$$

Due to (6.7), we see that (6.6c) becomes

$$\Theta_{|_0} = -\sigma^2 - \frac{1}{2}\Theta^2 + \Psi\Theta + \Phi^2\left\{b_j b^j + b^j{}_{|_j} + \Phi^2\omega^2\right\} - R_{00}. \tag{6.8}$$

According to the usual terminology, we call (6.8) the *Raychaudhuri equation* for the $(1+3)$ threading formalism determined by the non-normalized timelike vector field $\xi = \partial/\partial x^0$.

Next, we express the local components of the stress-energy-momentum tensor T with respect to a threading frame field, in terms of the quantities measured by an observer moving with unit 4-velocity

$$\mathbf{u} = \Phi^{-1}\frac{\partial}{\partial x^0}.$$

First, we note that

$$\rho = T(\mathbf{u}, \mathbf{u}), \tag{6.9}$$

is the relativistic energy density measured by the observer. Then, we put:

$$\begin{aligned} &(a)\quad T_{ij} = T\left(\frac{\delta}{\delta x^i}, \frac{\delta}{\delta x^j}\right), \quad (b)\quad T_{i0} = T\left(\frac{\delta}{\delta x^i}, \frac{\partial}{\partial x^0}\right), \\ &(c)\quad T_{00} = T\left(\frac{\partial}{\partial x^0}, \frac{\partial}{\partial x^0}\right), \end{aligned} \tag{6.10}$$

and define

$$(a)\quad p = \frac{1}{3}T_{ij}h^{ij}, \quad (b)\quad q_i = -\Phi^{-1}T_{i0}, \quad (c)\quad \pi_{ij} = T_{ij} - ph_{ij}. \tag{6.11}$$

Note that q_i and π_{ij} define spatial tensor fields of types $(0, 1)$ and $(0, 2)$, respectively. Moreover, comparing with the quantities defined on [4, p. 92], it is easy to see that p is the relativistic pressure, while q_i and π_{ij} determine completely the relativistic momentum density and the relativistic anisotropic (trace-free) stress tensor field, respectively. Finally, we conclude that the local components of the stress-energy-momentum tensor field T with respect to the threading frame field $\{\partial/\partial x^0, \delta/\delta x^i\}$, are given by

$$(a)\quad T_{ij} = \pi_{ij} + ph_{ij}, \quad (b)\quad T_{i0} = -\Phi q_i, \quad (c)\quad T_{00} = \Phi^2\rho. \tag{6.12}$$

7 The (1 + 3) Decomposition of Einstein Field Equations

Based on the (1 + 3) decomposition of both the Ricci tensor field and the stress-energy-momentum tensor field from the previous section, we express in a simple and elegant form the Einstein Field Equations (EFE).

Let the EFE given by (cf. [4, p. 65]):

$$Ric = 8\pi G(T - \frac{1}{2}\mathbf{T}g) + \Lambda g, \tag{7.1}$$

where G is the Newton constant, Λ is the cosmological constant, and $\mathbf{T}$ is the trace of T. Then, by applying the tensor fields to pairs of vector fields from the threading frame field $\{\partial/\partial x^0, \delta/\delta x^i\}$, and using (6.12), we deduce that (7.1) is equivalent to

$$\begin{aligned}
&(a)\quad R_{ik} = \{4\pi G(\rho - p) + \Lambda\}h_{ik} + 8\pi G\pi_{ik},\\
&(b)\quad R_{i0} = -8\pi G\Phi q_i,\\
&(c)\quad R_{00} = \Phi^2\{4\pi G(\rho + 3p) - \Lambda\}.
\end{aligned} \tag{7.2}$$

Finally, using (6.6a), (6.6b) and (6.8) into (7.2), we obtain

$$\begin{aligned}
(a)\quad & R^\star_{ik} - b_i b_k - \frac{1}{2}(b_{i|k} + b_{k|i}) + \Phi^{-2}\{\Theta_{ik|0} + (\Theta - \Psi)\Theta_{ik}\}\\
& \quad - \{4\pi G(\rho - p) + \Lambda\}h_{ik} - 8\pi G\pi_{ik} = 0,\\
(b)\quad & \Theta^j_{i|j} - \Theta_{|i} + \Theta c_i - \Theta^j_i c_j + \Phi^2\{\omega^j_{i|j} + \omega^j_i c_j + 2\omega^j_i b_j\}\\
& \quad + 8\pi G\Phi q_i = 0,\\
(c)\quad & \Theta_{|0} + \sigma^2 + \frac{1}{3}\Theta^2 - \Psi\Theta\\
& \quad - \Phi^2\{b^j_{\ |j} + b^2 + \Phi^2\omega^2 + 4\pi G(\rho + 3p) - \Lambda\} = 0,
\end{aligned} \tag{7.3}$$

where we put

$$b^2 = b_j b^j.$$

In spite of the huge literature on the $(1+3)$ threading of spacetime (cf. [4]), the Eq. (7.3), as far as we know, are stated here for the first time. This is because these equations apply for any Lorentz metric regardless the threading vector field and the integrability of the spatial distribution. Most of the literature on this matter presented these equations in the case of a unit vector field ξ which is also hypersurface orthogonal. This particular case applies to the Friedmann–Lemaître–Robertson–Walker universe, but it fails in any attempt to study metrics given by (2.7), with $\Phi \neq 0$ and $\omega \neq 0$. Such metrics are specific to both the black holes theory (cf. [3, 5]) and perturbation theory (cf. [7, 11]) where the splitting of the EFE given by (7.3) can be easy handled into the study.

References

1. Bejancu, A.: A new point of view on $(1+3)$ threading of spacetime. Adv. Math.: Sci. J. **4**(1), 21–42 (2015)
2. Bejancu, A., Călin, C.: On the (1+3) threading of spacetime with respect to an arbitrary timelike vector field. Eur. Phys. J. C **75**, 159 (2015)
3. Chandrasekhar, S.: The Mathematical Theory of Black Holes. Clarendon Press, Oxford (1983)
4. Ellis, G.F.R., Maartens, R., MacCallum, M.A.H.: Relativistic Cosmology. Cambridge University Press, Cambridge (2012)
5. Frolov, V.P., Novikov, I.D.: Black Hole Physics. Basic Concepts and New Developments. Kluwer, Dordrecht (1998)
6. Jantzen, R.T., Carini, P., Bini, D.: The many faces of gravitoelectromagnetism. Ann. Phys.-New York **215**, 1–50 (1992)
7. Kodama, H., Sasaki, M.: Cosmological perturbation theory. Prog. Theor. Phys. Supp. **78**, 1–166 (1984)
8. Maartens, R.: Linearization instability of gravity waves? Phys. Rev. D **55**, 463–467 (1997)
9. Mashhoon, B., McClune, J.C., Quevedo, H.: On the gravitoelectromagnetic stress-energy tensor. Classical Quant. Grav. **16**, 1137–1149 (1999)
10. Massa, E.: Space Tensors in General Relativity III. General Relativ. Gravit. **11**, 715–736 (1974)
11. Mukhanov, V.F., Feldman, H.A., Brandenberger, R.H.: Theory of cosmological perturbations. Phys. Rep. **215**, 203–333 (1992)
12. O'Neill, B.: Semi-Riemannian Geometry and Applications to Relativity. Academic Press, New York (1983)

Local Structure of Self-Dual Gradient Yamabe Solitons

Miguel Brozos-Vázquez, Esteban Calviño-Louzao, Eduardo García-Río and Ramón Vázquez-Lorenzo

Dedicated to Jaime Muñoz Masqué on the occasion of his 65th birthday

Abstract We analyze the underlying structure of a pseudo-Riemannian manifold admitting a gradient Yamabe soliton. Special attention is paid to neutral signature, where a description of self-dual gradient Yamabe solitons is obtained.

Keywords Yamabe soliton · Self-dual metric · Walker structure · Riemannian extension

1 Introduction

One of the central questions in modern Differential Geometry is the existence of canonical metrics on a given manifold M. Inspired by the uniformization theorem in dimension two, R.S. Hamilton proposed an approach to this problem based on

M. Brozos-Vázquez
Departamento de Matemáticas, Escola Politécnica Superior, Universidade da Coruña, Coruña, Spain
e-mail: miguel.brozos.vazquez@udc.es

E. Calviño-Louzao
Departamento de Matemáticas, IES Ramón Caamaño, Muxía, Spain
e-mail: estebcl@edu.xunta.es

E. García-Río (✉)
Facultade de Matemáticas, Universidade de Santiago de Compostela, 15782 Santiago de Compostela, Spain
e-mail: eduardo.garcia.rio@usc.es

R. Vázquez-Lorenzo
Departamento de Matemáticas, IES de Ribadeo Dionisio Gamallo, Ribadeo, Spain
e-mail: ravazlor@edu.xunta.es

M. Castrillón López et al. (eds.), *Geometry, Algebra and Applications: From Mechanics to Cryptography*, Springer Proceedings in Mathematics & Statistics 161, DOI 10.1007/978-3-319-32085-4_3

parabolic partial differential equations. In general, the goal of geometric evolution equations is to improve a given metric by considering the flow associated to a certain geometric object. Geometric flows have been applied to a variety of geometric, topological, and physical problems. The Ricci flow is one of the most extensively studied examples in the literature. One of its interests relies on the fact that, under appropriate conditions, it evolves the initial metric to an Einstein one. Other examples of geometric flows include the mean and the inverse mean curvature flows of submanifolds, the Kähler-Ricci and Calabi flows of manifolds, and the Yamabe and other conformal flows of metrics.

The *Yamabe flow* is an intrinsic geometric flow which can be interpreted as deforming a Riemannian metric to a conformal one of constant scalar curvature. It was introduced by Hamilton, shortly after the Ricci flow, as an approach to solve the Yamabe problem on manifolds of positive conformal Yamabe invariant. Formally, the Yamabe flow is defined by the evolution equation

$$\frac{\partial}{\partial t}g(t) = -\tau(t)g(t), \tag{1}$$

where $\tau(t)$ denotes the scalar curvature of $g(t)$. The Ricci flow and the Yamabe flow are equivalent in dimension $n = 2$, but they are essentially different in higher dimensions [2]. In fact, while the Yamabe flow leaves the conformal class of $g(0)$ invariant, the Ricci flow generically deforms $g(0)$ to a different conformal class. Hamilton proved that the Yamabe flow has a global solution for every initial metric [14], and conjectured that for any compact Riemannian manifold the unique solution of the Yamabe flow converges to a metric of constant scalar curvature. Hamilton's conjecture was proven by Ye [24] in the locally conformally flat case (see [2] for more information).

The genuine fixed points of the Yamabe flow are the metrics with zero scalar curvature. However, there are other kinds of self-similar solutions that have received attention in recent years, since they appear as possible singularity models. A family of metrics $g(t) = \sigma(t)\psi_t{}^*g(0)$ solving (1), where $\sigma(t)$ is a positive smooth function and $\psi_t : M \to M$ is a one-parameter family of diffeomorphisms of M, is said to be a *self-similar solution* of the Yamabe flow.

Self-similar solutions of (1) are in one to one correspondence with Yamabe solitons. As a matter of notation, a *Yamabe soliton* is a triple (M, g, X), where (M, g) is a pseudo-Riemannian manifold and X is a vector field on M satisfying

$$\mathcal{L}_X g = (\tau - \lambda)g. \tag{2}$$

Here τ denotes the scalar curvature of (M, g), $\mathcal{L}$ is the Lie derivative and $\lambda \in \mathbb{R}$. Whenever the vector field X is the gradient of a potential function f, we say that (M, g, f) is a *gradient Yamabe soliton* and for $X = \frac{1}{2}\nabla f$, Eq. (2) reduces to

$$\mathrm{Hess}_f = (\tau - \lambda)g, \tag{3}$$

where $\mathrm{Hess}_f = \nabla df$ is the Hessian tensor of f and ∇ denotes the Levi-Civita connection of (M, g).

It is important to emphasize that although the Yamabe flow is well-posed in the Riemannian setting, the existence of (even short-time) solutions is not guaranteed in the pseudo-Riemannian setting due to the lack of parabolicity of (1). However, the existence of self-similar solutions of the flow is equivalent to the existence of Yamabe solitons as in (2) (see [11]).

The main purpose of this work is to determine the local structure of self-dual gradient Yamabe solitons (see Theorem 4). As an application, we show the existence of self-dual gradient Yamabe solitons in neutral signature that are not locally conformally flat (see Remarks 5 and 6).

2 Local Structure of Pseudo-Riemannian Gradient Yamabe Solitons

We study the local structure of a pseudo-Riemannian gradient Yamabe soliton (M, g, f) in general, without restrictions on the dimension or the signature, unless specification on the contrary. We emphasize that the gradient Yamabe soliton equation (3) codifies geometric information about the structure of (M, g) by means of the scalar curvature and the second fundamental form of the level sets of the potential function f. The analysis of the level sets of f naturally splits into two different cases. The first case corresponds to non-degenerate hypersurfaces (i.e., $\|\nabla f\| \neq 0$), and (M, g, f) is called a *non-isotropic* gradient Yamabe soliton. The second case corresponds to degenerate hypersurfaces (i.e., $\|\nabla f\| = 0$ but $\nabla f \neq 0$), and (M, g, f) is called an *isotropic* gradient Yamabe soliton. We analyze both cases separately in what follows. When f is constant the Yamabe soliton is said to be *trivial* and will be excluded from our analysis.

By contracting Eq. (3) one immediately obtains that $(1/n)\Delta f = \tau - \lambda$. This relation shows that the Yamabe soliton equation is a special case of the more general Möbius equation

$$\mathrm{Hess}_f = \frac{\Delta f}{n} g. \tag{4}$$

As an application of (4) one gets that the level sets of the function f are totally umbilical and the normalized gradient vector field $\nabla f / \|\nabla f\|$ is a non-null geodesic vector field. Hence the integral curves of ∇f are (unparametrized) geodesics and it follows from [20] that (M, g) has the local structure of a twisted product. Furthermore one shows that the level sets of f are indeed spherical hypersurfaces to get that the underlying structure of any gradient non-isotropic Yamabe soliton is a warped product. (See [13, 15, 16] for a discussion of the local structure of pseudo-Riemannian manifolds admitting a closed conformal vector field, and [22] for the local structure of solutions of the Möbius equation).

Theorem 1 *Let (M, g) be a pseudo-Riemannian manifold. If (M, g, f) is a non-isotropic gradient Yamabe soliton then (M, g) is locally isometric to a warped product $(-\varepsilon, \varepsilon) \times_{\varphi(t)} N$, where (N, g_N) is a pseudo-Riemannian manifold with constant scalar curvature.*

The potential function f of the soliton is related to the warping function φ. In a neighborhood of a regular point of f one has that $\varphi = f'$, while a different relation holds in a neighborhood of a critical point of f (see [15, 16]).

The local warped product decomposition in the previous theorem can be more precise in a neighborhood of a critical point of the soliton function f. In such a case the fiber (N, g_N) is of constant curvature so that (M, g) is locally conformally flat [15, 16].

Remark 1 In Riemannian signature a global structure result was given in [10] showing that the local warped product decomposition is global in the complete Riemannian setting.

Observe that if (M, g, f) is a gradient Yamabe soliton, then ∇f is a conformal vector field. Moreover, if the scalar curvature is constant, then ∇f is a homothetic vector field on (M, g). Hence, in this case, either $\nabla f = 0$ or (M, g) is flat in the constant scalar curvature Riemannian setting.

Remark 2 An immediate application of the Theorem 1 shows that any pseudo-Riemannian self-dual gradient Yamabe soliton is necessarily locally conformally flat, since self-dual warped products are locally conformally flat [5].

Next, let (M, g, f) be an isotropic gradient Yamabe soliton (so $\|\nabla f\| = 0$ on an open subset $\mathcal{U} \subset M$). Then we have the following restrictions on the geometry of (M, g).

Theorem 2 *Let (M, g, f) be an isotropic pseudo-Riemannian gradient Yamabe soliton. Then the scalar curvature of (M, g) is constant $\tau = \lambda$ and ∇f is a parallel null vector field.*

Proof Since $\|\nabla f\| = 0$, it follows that

$$\mathrm{Hess}_f(X, \nabla f) = g(\nabla_X \nabla f, \nabla f) = \frac{1}{2} X g(\nabla f, \nabla f) = 0$$

for any vector field X. Hence the soliton equation (3) implies $\tau = \lambda$, thus showing that the scalar curvature is constant and, moreover, that ∇f is a parallel null vector field. □

Remark 3 Theorems 1 and 2 immediately apply to Lorentzian gradient Yamabe solitons to provide a description of the local structure. Recall that a Lorentzian manifold is called a *Brinkmann wave* if it admits a parallel null vector field X. In such a case (see [3]), there exist coordinates $(u, v, x^1, \ldots, x^{n-2})$ so that

$$g = 2du \circ dv + \sum_{i,j=1}^{n-2} g_{ij}\, dx^i \circ dx^j \,, \tag{5}$$

where $\partial g_{ij}/\partial v = 0$ and the parallel null vector field is $X = \partial_u$. Hence $X = \nabla v$. In conclusion, if (M, g, f) is a non-trivial Lorentzian gradient Yamabe soliton then if it is:

(i) non-isotropic, then (M, g) is locally isometric to a warped product $(-\varepsilon, \varepsilon) \times_{\varphi(t)} N$, where (N, g_N) is a pseudo-Riemannian manifold with constant scalar curvature, and
(ii) isotropic, then (M, g) is locally isometric to a Brinkmann wave with coordinates as in (5) and the potential function f is given by the coordinate function $f = v$.

Differently to the Riemannian case, Lorentzian gradient Yamabe solitons of constant scalar curvature are not necessarily flat. Indeed, the existence of non-Killing homothetic vector fields is not so restrictive in the strictly pseudo-Riemannian setting as it is in the Riemannian one. This is due to the existence of pseudo-Riemannian manifolds with non-zero nilpotent Ricci operators (observe that any homothetic vector field is a Ricci collineation). See [9] for a classification of three-dimensional homogeneous Lorentzian Yamabe solitons.

3 Self-dual Gradient Yamabe Solitons: Local Structure and Examples

In this section we study self-dual gradient Yamabe solitons in dimension four. Since any self-dual warped product metric is locally conformally flat [5], it follows from Theorem 1 that any non-isotropic self-dual gradient Yamabe soliton is locally conformally flat. Moreover, any self-dual Lorentzian metric is locally conformally flat, therefore in what follows we restrict to isotropic Yamabe solitons in neutral signature $(--++)$. Our main result not only shows the existence of self-dual gradient Yamabe solitons which are not locally conformally flat, but also provides the local structure of the underlying manifold as we shall see in Theorem 4.

Let (M, g, f) be a non-trivial isotropic gradient Yamabe soliton with pseudo-Riemannian metric of neutral signature $(--++)$. Since ∇f is non-zero and null, there exist a unit timelike vector field e_1 and a unit spacelike vector field e_3 such that $\nabla f = (1/\sqrt{2})(e_1 + e_3)$.

Complete $\{e_1, e_3\}$ to a local orthonormal frame $\{e_1(-), e_2(-), e_3(+), e_4(+)\}$, where $(\pm)$ indicates the causal character of e_i. Consider now the locally defined vector fields

$$\nabla f = \frac{1}{\sqrt{2}}(e_1 + e_3), \qquad U = \frac{1}{\sqrt{2}}(e_2 + e_4),$$

and observe that $\mathscr{D} = \text{span}\{\nabla f, U\}$ is a totally isotropic distribution which is defined locally.

As a matter of terminology, a pseudo-Riemannian manifold (M, g) is said to be a *Walker manifold* if it admits a parallel null distribution (see [6] for more information on Walker structures). We have the following result.

Lemma 1 *Let (M, g, f) be a non-trivial isotropic gradient Yamabe soliton with g of signature $(--++)$. Then the null distribution $\mathscr{D} = \text{span}\{\nabla f, U\}$ is parallel and hence (M, g) is a Walker manifold.*

Proof First of all observe from Theorem 2 that the scalar curvature of (M, g) is constant $\tau = \lambda$ and ∇f is a parallel null vector field. Now we consider the null distribution $\mathscr{D} = \text{span}\{\nabla f, U\}$ and show that it is parallel as follows. Since $\mathscr{D}$ is totally isotropic and $\mathscr{D}^{\perp} = \mathscr{D}$ we see that $\nabla_X U \in \mathscr{D}$:

$$\begin{aligned} 0 &= X\, g(U, U) = 2g(\nabla_X U, U), \\ 0 &= X\, g(U, \nabla f) = g(\nabla_X U, \nabla f) + g(U, \nabla_X \nabla f) = g(\nabla_X U, \nabla f). \end{aligned}$$

Therefore $\nabla_X \mathscr{D} \subset \mathscr{D}$ for all vector fields X, which shows that $\mathscr{D}$ is parallel. □

Self-dual Walker metrics are locally isometric to cotangent bundles over affine surfaces [8]. In order to describe such metrics we briefly recall the following terminology about the geometry of cotangent bundles and refer to [23] (see also [6]) for further details. Let T^*M be the cotangent bundle of an n-dimensional manifold M and let $\pi : T^*M \to M$ be the projection. Let $\tilde{p} = (p, \omega)$, where $p \in M$ and $\omega \in \bigwedge^1(T_pM)$, denote a point of T^*M. Local coordinates (x^i) in a neighborhood $\mathscr{U}$ in M induce coordinates $(x^i, x_{i'})$ in $\pi^{-1}(\mathscr{U})$, where ω decomposes as $\omega = \sum x_{i'} dx^i$.

For each vector field X on M, define a function $\iota X : T^*M \to \mathbb{R}$ by $\iota X(p, \omega) = \omega(X_p)$. Expand $X = X^j \partial_j$ and express $\iota X(x_i, x_{i'}) = \sum x_{i'} X^i$. The importance of the evaluation functions ιX relies on vector fields on T^*M being completely determined by their action on evaluations: if $\tilde{Y}$, $\tilde{Z}$ are vector fields on T^*M, then $\tilde{Y} = \tilde{Z}$ if and only if $\tilde{Y}(\iota X) = \tilde{Z}(\iota X)$ for all vector fields X on M. The above result allows the lifting construction: for any vector field X on M, its *complete lift* X^C is the vector field on T^*M characterized by the identity $X^C(\iota Z) = \iota[X, Z]$ for all vector fields Z on M. The main significance of complete lifts of vector fields is that $(0, s)$-tensor fields on T^*M are determined by their action on complete lifts.

Next, let D be a torsion-free affine connection on M. The *Riemannian extension* g_D is the pseudo-Riemannian metric g_D on T^*M of neutral signature (n, n) characterized by the identity $g_D(X^C, Y^C) = -\iota(\nabla_X Y + \nabla_Y X)$. In order to express locally the Riemannian extension, let $D_{\partial_i} \partial_j = {}^D\Gamma_{ij}{}^k \partial_k$ give the Christoffel symbols of the connection D. Then:

$$g_D = 2\, dx^i \circ dx_{i'} - 2 x_{k'}\, {}^D\Gamma_{ij}{}^k dx^i \circ dx^j.$$

Riemannian extensions were originally defined by Patterson and Walker [19] and further investigated in [1], showing the relation between pseudo-Riemannian properties of T^*M with the affine structure of the base manifold (M, D). Let Φ be a symmetric $(0, 2)$-tensor field on M. The *deformed Riemannian extension* is the neutral signature metric on T^*M given by $g_{D,\Phi} = g_D + \pi^*\Phi$.

Let T be a tensor field of type $(1, 1)$ on M and define a 1-form ιT on T^*M which is characterized by the identity $\iota T(X^C) = \iota(TX)$. The *modified Riemannian extension* is the neutral signature metric on T^*M defined by

$$g_{D,\Phi,T,S} := \iota T \circ \iota S + g_D + \pi^*\Phi,$$

where T and S are $(1, 1)$-tensor fields on M and Φ is a symmetric $(0, 2)$-tensor field on M. In a system of induced local coordinates one has

$$g_{D,\Phi,T,S} = 2\,dx^i \circ dx_{i'} + \left\{\tfrac{1}{2}x_{r'}x_{s'}\left(T_i^r S_j^s + T_j^r S_i^s\right) + \Phi_{ij}(x) - 2x_{k'}{}^D\Gamma_{ij}{}^k\right\} dx^i \circ dx^j. \tag{6}$$

Self-dual Walker metrics are obtained by a deformation of the modified Riemannian extensions above. The following result gives a local description of self-dual Walker metrics and provides a large family of examples of non-locally conformally flat self-dual metrics.

Theorem 3 ([8]) *A Walker metric in signature $(--++)$ is self-dual if and only if it is locally isometric to the cotangent bundle $T^*\Sigma$ of an affine surface (Σ, D), with metric tensor*

$$g_{D,\Phi,T,\mathrm{id},X} = \iota X(\iota\mathrm{id} \circ \iota\mathrm{id}) + \iota\mathrm{id} \circ \iota T + g_D + \pi^*\Phi,$$

where D, Φ, T and X are a torsion-free affine connection, a symmetric $(0, 2)$-tensor field on Σ, a $(1, 1)$-tensor field and a vector field, respectively.

As an application of the previous result and Lemma 1, any isotropic self-dual gradient Yamabe soliton is a modified Riemannian extension $g_{D,\Phi,T,X}$. Moreover, one has

Theorem 4 *Let (M, g, f) be a non-trivial self-dual isotropic gradient Yamabe soliton of neutral signature $(--++)$. Then (M, g) is locally isometric to the cotangent bundle $T^*\Sigma$ of an affine surface (Σ, D) equipped with a deformed Riemannian extension $g_{D,\Phi}$. Furthermore, the potential function of the soliton is of the form $f = \tilde{f} \circ \pi$, for some function $\tilde{f}$ on Σ which is affine, i.e., $Dd\tilde{f} = 0$.*

Proof By Lemma 1 (M, g) is a Walker manifold and $\nabla f \in \mathcal{D}$, where $\mathcal{D}$ is the parallel null distribution. By Theorem 3, (M, g) is locally isometric to the cotangent bundle $T^*\Sigma$ of an affine surface and, moreover, the null distribution satisfies $\mathcal{D} = \ker \pi_*$ (see [8]). Now observe that the scalar curvature of the modified Riemannian extensions $g_{D,\Phi,T,\mathrm{id},X}$ at Theorem 3 is given by $\tau = 12\iota X + 3\mathrm{trace}(T)$. Hence, since the scalar curvature of any isotropic gradient Yamabe soliton is constant $\tau = \lambda$, it immediately follows that the vector field X vanishes identically and, moreover, that $\mathrm{trace}(T) = \frac{1}{3}\lambda$.

Let $f : T^*\Sigma \to \mathbb{R}$ be the potential function of the soliton. Since $\mathrm{Hess}_f = 0$, considering the components

$$\mathrm{Hess}_f(\partial_{x_{i'}}, \partial_{x_{j'}}) = \frac{\partial^2}{\partial x_{i'} \partial x_{j'}} f$$

it follows that the function f is of the form

$$f = \iota Z + (\tilde{f} \circ \pi) \tag{7}$$

for some vector field Z on Σ and some function $\tilde{f}\colon \Sigma \to \mathbb{R}$.

Assuming the vector field Z is not identically zero, choose coordinates on Σ so that $Z = \partial_{x^1}$. Since the potential function f is expressed as $f(x^1, x^2, x_{1'}, x_{2'}) = x_{1'} + \tilde{f}(x^1, x^2)$, its gradient becomes

$$\nabla f = \partial_{x^1} + (\partial_{x^1}\tilde{f} - \Phi_{11})\partial_{x_{1'}} + (\partial_{x^2}\tilde{f} - \Phi_{12})\partial_{x_{2'}},$$

which is a contradiction since ∇f should belong to the parallel null distribution $\mathscr{D} = \ker\pi_* = \mathrm{span}\{\partial_{x_{1'}}, \partial_{x_{2'}}\}$.

Hence assume in what follows that the vector field Z is identically zero. Then (7) reduces to $f(x^1, x^2, x_{1'}, x_{2'}) = \tilde{f}(x^1, x^2)$ and the modified Riemannian extension $g_{D,\Phi,T,\mathrm{id}}$ is locally expressed as

$$g_{D,\Phi,T,\mathrm{id}} = 2\,dx^i \circ dx_{i'} + \left\{\tfrac{1}{2}x_{r'}x_{s'}\left(T_i^r\delta_j^s + T_j^r\delta_i^s\right) + \Phi_{ij}(x) - 2x_{k'}{}^D\Gamma_{ij}{}^k\right\} dx^i \circ dx^j.$$

Note from [12] that one may choose appropriate coordinates on Σ so that ${}^D\Gamma_{11}^\ell = 0$, $\ell = 1, 2$. Now, a long but straightforward calculation shows that the only non-zero components of Hess_f are

$$\begin{aligned}
\mathrm{Hess}_f(\partial_{x^1}, \partial_{x^1}) &= \tilde{f}_{11} + x_{1'}T_1^1\tilde{f}_1 + \tfrac{1}{2}T_1^2(x_{1'}\tilde{f}_2 + x_{2'}\tilde{f}_1),\\
\mathrm{Hess}_f(\partial_{x^2}, \partial_{x^2}) &= \tilde{f}_{22} + \tfrac{1}{2}\left(\left(x_{2'}T_2^1 - 2{}^D\Gamma_{22}^1\right)\tilde{f}_1 + \left(x_{1'}T_2^1 + 2x_{2'}T_2^2 - 2{}^D\Gamma_{22}^2\right)\tilde{f}_2\right),\\
\mathrm{Hess}_f(\partial_{x^1}, \partial_{x^2}) &= \tilde{f}_{12} + \tfrac{1}{4}\left(\left(2x_{2'}T_1^2 + x_{1'}\left(T_1^1 + T_2^2\right) - 4{}^D\Gamma_{12}^2\right)\tilde{f}_2\right.\\
&\quad \left. + \left(2x_{1'}T_2^1 + x_{2'}\left(T_1^1 + T_2^2\right) - 4{}^D\Gamma_{12}^1\right)\tilde{f}_1\right),
\end{aligned}$$

where the $\tilde{f}_{ij} = \partial^2\tilde{f}/\partial x^i\partial x^j$ denote the partial derivatives of $\tilde{f}$, and the scalar curvature satisfies $\tau = 3(T_1^1 + T_2^2) = \lambda$.

A straightforward calculation from the previous equations (since the potential function f is not constant) shows that the $(1, 1)$-tensor field T vanishes identically (and so the scalar curvature and the soliton constant λ do). Hence the metric reduces to a Riemannian extension $g_{D,\Phi}$ and the equations above become

$$\begin{aligned}
\mathrm{Hess}_f(\partial_{x^1}, \partial_{x^1}) &= \tilde{f}_{11}, \\
\mathrm{Hess}_f(\partial_{x^2}, \partial_{x^2}) &= \tilde{f}_{22} - {}^D\Gamma_{22}^1 \tilde{f}_1 - {}^D\Gamma_{22}^2 \tilde{f}_2, \\
\mathrm{Hess}_f(\partial_{x^1}, \partial_{x^2}) &= \tilde{f}_{12} - {}^D\Gamma_{12}^2 \tilde{f}_2 - {}^D\Gamma_{12}^1 \tilde{f}_1.
\end{aligned}$$

These show that $\mathrm{Hess}_f = 0$ if and only if $Dd\tilde{f} = 0$. □

Remark 4 If the Riemannian extension $g_{D,\Phi}$ is locally conformally flat, then the connection D must be projectively flat [1]. Therefore, if (Σ, D) is a non-projectively flat surface, then the isotropic gradient Yamabe solitons constructed in Theorem 4 are self-dual but not locally conformally flat, in contrast with the non-isotropic case.

Furthermore observe that the construction in Theorem 4 is independent of the symmetric (0, 2)-tensor field Φ. Therefore, given any affine function $\tilde{f}$ on (Σ, D), its pull-back $f = \tilde{f} \circ \pi$ defines an isotropic Yamabe soliton on $(T^*\Sigma, g_{D,\Phi})$ for any symmetric (0, 2)-tensor field Φ.

Remark 5 Following the discussion in [4], one can obtain examples of affine surfaces (Σ, D) admitting non-constant affine functions $\tilde{f}$ as follows. Let h be a solution of the affine gradient Ricci soliton equation on (Σ, D) (i.e., $Ddh + \rho^D_{sym} = 0$, where ρ^D_{sym} is the symmetric part of the Ricci tensor of (Σ, D)). Then for any affine-Killing vector field ξ (i.e., $\mathscr{L}_\xi D = 0$) one has that $\tilde{f} = \xi(h)$ is an affine function and hence it defines an isotropic gradient Yamabe soliton on $T^*\Sigma$ (see also [7]).

Examples of affine gradient Ricci solitons can be constructed on homogeneous affine surfaces as follows. First of all recall from [18] that if D is a homogeneous affine connection, then it is the Levi-Civita connection of a metric of constant Gauss curvature or, otherwise, there are coordinates (x^1, x^2) such that

(A) all Christoffel symbols are constants, i.e., ${}^D\Gamma_{ij}^k = \gamma_{ij}^k$, $\gamma_{ij}^k \in \mathbb{R}$, or
(B) all Christoffel symbols are of the form ${}^D\Gamma_{ij}^k = (1/x^1)\gamma_{ij}^k$, $\gamma_{ij}^k \in \mathbb{R}$.

The existence of affine Ricci solitons on projectively flat surfaces implies that the Ricci tensor is degenerate, and hence D is flat if it is the Levi-Civita connection of a surface of constant curvature. By contrast, affine connections as in (A) admit non-trivial affine gradient Ricci solitons if and only if the Ricci tensor is of rank one. Connections of type (B) also admit non-trivial affine gradien Ricci solitons as shown in [7].

Remark 6 Let D be an affine connection with symmetric Ricci tensor of rank one and such that the kernel of ρ^D is parallel. There exist adapted coordinates (x^1, x^2) where the only non-zero Christoffel symbols are [17]

$$ {}^D\Gamma_{12}^1 \quad \text{and} \quad {}^D\Gamma_{22}^1, \quad \text{where} \quad \partial_{x^1} {}^D\Gamma_{12}^1 = 0. \tag{8}$$

Moreover, it follows that the only non-zero component of the Ricci tensor is given by $\rho^D(\partial_{x^2}, \partial_{x^2}) = \partial_{x^1} {}^D\Gamma_{22}^1 - \partial_{x^2} {}^D\Gamma_{12}^1 - ({}^D\Gamma_{12}^1)^2$ and the connection D is projectively flat if and only if $\partial^2_{x^1x^1} {}^D\Gamma_{22}^1 = 0$. A straightforward calculation shows that a function $\tilde{f}(x^1, x^2)$ is affine if and only if (see [4])

$$(Dd\tilde{f})(\partial_{x^1},\partial_{x^1}) = \tilde{f}_{11}, \quad (Dd\tilde{f})(\partial_{x^1},\partial_{x^2}) = \tilde{f}_{12} - {}^D\Gamma_{12}^1\tilde{f}_1,$$
$$(Dd\tilde{f})(\partial_{x^2},\partial_{x^2}) = \tilde{f}_{22} - {}^D\Gamma_{22}^1\tilde{f}_1,$$

so $\tilde{f}(x^1,x^2) = x^1h(x^2) + \hat{h}(x^2)$. Then the above equations reduce to

$$h'(x^2) = {}^D\Gamma_{12}^1(x^2)h(x^2), \qquad \hat{h}''(x^2) = h(x^2)^D\Gamma_{22}^1(x^1,x^2) - x^1h''(x^2).$$

Working with the previous equations and the expression of the Ricci tensor, one has that the compatibility condition $(\partial_{x^1}(h(x^2)^D\Gamma_{22}^1(x^1,x^2) - x^1h''(x^2)) = 0)$ reduces to $h(x^2)\rho^D(\partial_{x^2},\partial_{x^2}) = 0$. This shows that $h(x^2)$ vanishes identically in a neighborhood of any point where D is non-flat. Hence the potential function of the soliton is of the form $\tilde{f}(x^1,x^2) = \alpha x^2 + \beta$ if the connection D is non-flat ($\alpha, \beta \in \mathbb{R}$).

If the connection D is flat (in which case ${}^D\Gamma_{22}^1(x^1,x^2) = x^1({}^D\Gamma_{12}^1(x^2) + {}^D\Gamma'(x^2)) + \mathscr{A}(x^2)$ for some function $\mathscr{A}(x^2)$), then

$$\tilde{f}(x^1,x^2) = x^1h(x^2) + \hat{h}(x^2),$$

where

$$h'(x^2) = {}^D\Gamma_{12}^1(x^2)h(x^2), \qquad \hat{h}''(x^2) = h(x^2)\mathscr{A}(x^2).$$

In any of these cases the Riemannian extension results in a self-dual isotropic gradient Yamabe soliton $(T^*\Sigma, g_{D,\Phi}, \tilde{f}\circ\pi)$, which is not locally conformally flat for generic Φ.

4 Conclusion

The analysis of the local structure of gradient Yamabe solitons shows that any non-isotropic (i.e., $\|\nabla f\| \neq 0$) self-dual gradient Yamabe soliton (M, g, f) is locally conformally flat. However isotropic examples (i.e., $\|\nabla f\| = 0$) that are not locally conformally flat exist in the neutral signature case and they are realized on the cotangent bundle of affine surfaces that admit non-constant affine functions.

Acknowledgments This work has been supported by projects GRC2013-045, MTM2013-41335-P and EM2014/009 with FEDER funds (Spain).

References

1. Afifi, Z.: Riemann extensions of affine connected spaces. Q. J. Math. **5**, 312–320 (1954)
2. Brendle, S.: Evolution equations in Riemannian geometry. Jpn. J. Math. **6**, 45–61 (2011)
3. Brinkmann, H.W.: Einstein spaces which are mapped conformally on each other. Math. Ann. **94**, 119–145 (1925)

4. Brozos-Vázquez, M., García-Río, E.: Four-dimensional neutral signature self-dual gradient Ricci soliton. Indiana Univ. Math. J. (to appear)
5. Brozos-Vázquez, M., García-Río, E., Vázquez-Lorenzo, R.: Osserman and conformally Osserman manifolds with warped and twisted product structure. Results Math. **52**, 211–221 (2008)
6. Brozos-Vázquez, M., García-Río, E., Gilkey, P., Nikčević, S., Vázquez-Lorenzo, R.: The Geometry of Walker Manifolds. Synthesis Lectures on Mathematics and Statistics, vol. 5. Morgan & Claypool Publ., Williston (2009)
7. Brozos-Vázquez, M., García-Río, E., Gilkey, P.: Homogeneous affine surfaces: Affine Killing vector fields and gradient Ricci solitons. arXiv:1512.05515 (2015)
8. Calviño-Louzao, E., García-Río, E., Gilkey, P., Vázquez-Lorenzo, R.: The geometry of modified Riemannian extensions. Proc. R. Soc. A-Math. Phys. **465**, 2023–2040 (2009)
9. Calviño-Louzao, E., Seoane-Bascoy, J., Vázquez-Abal, M.E., Vázquez-Lorenzo, R.: Three-dimensional homogeneous Lorentzian Yamabe solitons. Abh. Math. Sem. Hamburg **82**, 193–203 (2012)
10. Cao, H.-D., Sun, X., Zhang, Y.: On the structure of gradient Yamabe solitons. Math. Res. Lett. **19**, 767–774 (2012)
11. Chow, B., Lu, P., Ni, L.: Hamilton's Ricci Flow. Graduate Studies in Mathematics, vol. 77. American Mathematical Society, Providence (2006). Science Press, New York
12. Dusek, Z., Kowalski, O.: How many are affine connections with torsion. Arch. Math. (Brno) **50**, 257–264 (2014)
13. Fialkow, A.: Conformal geodesics. Trans. Am. Math. Soc. **45**, 443–473 (1939)
14. Hamilton, R.S.: Lectures on geometric flows (unpublished) (1989)
15. Kerbrat, Y.: Transformations conformes des variétés pseudo-Riemanniennes. J. Differ. Geom. **11**, 547–571 (1976)
16. Kühnel, W., Rademacher, H.-B.: Essential conformal fields in pseudo-Riemannian geometry. J. Math. Pure Appl. **74**, 453–481 (1995)
17. Opozda, B.: A class of projectively flat surfaces. Math. Z. **219**, 77–92 (1995)
18. Opozda, B.: A classification of locally homogeneous connections on 2-dimensional manifolds. Differ. Geom. Appl. **21**, 173–198 (2004)
19. Patterson, E.M., Walker, A.G.: Riemann extensions. Q. J. Math. **3**, 19–28 (1952)
20. Ponge, R., Reckziegel, H.: Twisted products in pseudo-Riemannian geometry. Geometriae Dedicata **48**, 15–25 (1993)
21. Walker, A.G.: Canonical form for a Riemannian space with a parallel field of null planes. Q. J. Math. **2**(1), 69–79 (1950)
22. Xu, X.: On the existence and uniqueness of solutions of Möbius equations. Trans. Am. Math. Soc. **337**, 927–945 (1993)
23. Yano, K., Ishihara, S.: Tangent and Cotangent Bundles. Marcel Dekker, New York (1973)
24. Ye, R.: Global existence and convergence of Yamabe flow. J. Differ. Geom. **39**, 35–50 (1994)

The Prescribed Curvature Problem in Low Dimension

Giovanni Calvaruso

Dedicated to Jaime Muñoz Masqué on the occasion of his 65th birthday

Abstract We describe some recent results concerning the inverse curvature problem, that is, the existence and description of metrics with prescribed curvature, focusing on the low-dimensional homogeneous cases.

Keywords Homogeneous Lorentzian metrics · Ricci curvature · Segre types

1 Introduction

Geometric properties of a pseudo-Riemannian manifold (M, g) are encoded in its curvature, and usually expressed by some conditions on the curvature tensor itself. Starting from the metric tensor g, the curvature tensor R of (M, g) can be completely determined. The inverse problem, namely, to determine a pseudo-Riemannian manifold with assigned curvature, is known as the *prescribed curvature problem*, and it has been extensively studied. In this framework, two distinct problems naturally arise:

(i) Existence results: necessary and sufficient conditions for an assigned two-form on a manifold to be (locally) the curvature form of a pseudo-Riemannian metric.
(ii) Explicit examples of such a metric.

G. Calvaruso (✉)
Department of Mathematics and Physics "Ennio De Giorgi", University of Salento, 73100 Lecce, LE, Italy
e-mail: giovanni.calvaruso@unisalento.it

M. Castrillón López et al. (eds.), *Geometry, Algebra and Applications: From Mechanics to Cryptography*, Springer Proceedings in Mathematics & Statistics 161, DOI 10.1007/978-3-319-32085-4_4

The study of the first problem led to local existence theorems under very general hypotheses (see for example [6–11, 14] and references therein). In particular, as proved by DeTurck [6–8], if a symmetric (0, 2)-tensor $\mathscr{R}$ is analytic in a neighborhood of a point $x_0 \in \mathbb{R}^n$ and $\mathscr{R}^{-1}(x_0)$ exists, then there exists an analytic metric g, of any desired signature, such that $\mathscr{R} = \rho$ is the Ricci tensor of g in a neighborhood of x_0. The *Bianchi identity*

$$\operatorname{Bian}(g, \mathscr{R}) = g^{ab}\left(\mathscr{R}_{am;b} - \tfrac{1}{2}\mathscr{R}_{ab;m}\right) = 0$$

yields some restrictions for the 2-forms admissible as curvature forms. It is worth to emphasize the physical meaning of such restrictions. In fact, $\operatorname{Bian}(g, \mathscr{R}) = -\operatorname{div}(G\mathscr{R})$, where G is the gravitation operator $Gh = h_{ij} - \frac{1}{2}g_{ij}(g^{ab}h_{ab})$. In particular, $G\rho$ is the stress-energy tensor in Einstein's theory of gravitation [6].

In this framework, low-dimensional cases have some special properties. In fact, in dimension three every 2-form with values in a semi-simple Lie algebra is generically the curvature of a connection form locally [9, 10, 14]. Moreover, in dimension four, Bianchi's identities can be eliminated for a large class of Lie algebras (which strictly includes the semi-simple ones). Curvature forms can be then characterized as the solutions to a second-order partial differential system, which was proved in [11] to be formally integrable.

On the other hand, even in special cases, as in low dimension and for particularly simple forms of the curvature or the Ricci tensor, the second problem is still open (up to our knowledge). Moreover, it is a natural problem to look for *homogeneous* metrics of prescribed curvature, since they are the homogeneous models for metrics of the same dimension. Also with regard to the existence problem, the above cited Refs. [9–11, 14] showed the special role played by homogeneous examples (in particular, Lie groups and the corresponding Lie algebras).

In this framework, the three-dimensional case acquires a peculiar relevance, for several reasons. First of all, in dimension three the Ricci tensor completely determines the curvature. Moreover, a connected, simply connected, complete three-dimensional homogeneous manifold is either symmetric or isometric to some Lie group equipped with a left-invariant metric (we may refer to [13] for the Riemannian case and [1] for the Lorentzian one). Finally, with the obvious exceptions of $\mathbb{R} \times \mathbb{S}^2$ (Riemannian) and $\mathbb{R}_1 \times \mathbb{S}^2$ (Lorentzian), three-dimensional connected simply connected symmetric spaces are also realized in terms of suitable left-invariant metrics on Lie groups [2].

In this note we will illustrate how three-dimensional locally homogeneous Lorentzian metrics on $\mathbb{R}^3$ were constructed in [3] for all admissible Ricci operators, that is, for all real-valued matrices which can occur as the Ricci operator of a homogeneous Lorentzian three-manifold. To do so, we introduce a system of partial differential equations, whose solutions determine explicitly these Lorentzian metrics. Then, solutions are presented for *proper* Lorentzian models, that is, Lorentzian homogeneous three-spaces which do not have any counterpart in Riemannian geometry, since their Ricci operator is not diagonalizable. We also mention the fact that explicit examples for the wider class of *curvature homogeneous* Lorentzian three-manifolds were constructed in [4, 5], proving that for all Segre types of the Ricci operator, there exist examples of curvature homogeneous Lorentzian metrics in $\mathbb{R}^3$.

2 Locally Homogeneous Lorentzian Three-Manifolds

Let (M, g) be a connected Lorentzian three-manifold. We denote by ∇ the Levi-Civita connection of (M, g) and by R its curvature tensor, taken with the sign convention $R(X, Y) = \nabla_{[X,Y]} - [\nabla_X, \nabla_Y]$. Since $\dim M = 3$, R is completely determined by the Ricci tensor ρ, defined by $\rho(X, Y)_p = \sum_{i=1}^{3} \varepsilon_i g(R(X, e_i)Y, e_i)$, where $\{e_i\}$ is a pseudo-orthonormal basis of T_pM and $\varepsilon_i = g(e_i, e_i) = \pm 1$ for all i. Throughout the paper we shall assume that e_3 is *timelike*, that is, $\varepsilon_1 = \varepsilon_2 = -\varepsilon_3 = 1$.

Because of the symmetries of R, the Ricci tensor ρ is symmetric. Consequently, the *Ricci operator* Q, defined by $g(QX, Y) = \rho(X, Y)$, is self-adjoint. Thus, in the Riemannian case there exists an orthonormal basis diagonalizing Q, while for a Lorentzian manifold there exists a suitable pseudo-orthonormal basis $\{e_1, e_2, e_3\}$, with e_3 timelike, such that Q takes one of the following forms, called *Segre types*:

$$\text{Segre type } \{11, 1\}: \begin{pmatrix} \bar a & 0 & 0 \\ 0 & \bar b & 0 \\ 0 & 0 & \bar c \end{pmatrix}, \qquad \text{Segre type } \{1z\bar z\}: \begin{pmatrix} \bar a & 0 & 0 \\ 0 & \bar b & \bar c \\ 0 & -\bar c & \bar b \end{pmatrix},$$

$$\text{Segre type } \{21\}: \begin{pmatrix} \bar a & 0 & 0 \\ 0 & \bar b & \varepsilon \\ 0 & -\varepsilon & \bar b - 2\varepsilon \end{pmatrix}, \qquad \text{Segre type } \{3\}: \begin{pmatrix} \bar b & \bar a & -\bar a \\ \bar a & \bar b & 0 \\ \bar a & 0 & \bar b \end{pmatrix}.$$

If (M, g) is curvature homogeneous (in particular, locally homogeneous), then its Ricci operator Q has the same Segre type at every point $p \in M$ and there exists (at least, locally) a pseudo-orthonormal frame field $\{e_i\}$ such that Q is given by one of the expressions above, for some constants $\bar a$, $\bar b$ and $\bar c$. As in [1], we now put

$$\nabla_{e_i} e_j = \sum_k \varepsilon_j b^i_{jk} e_k, \tag{1}$$

for all indices i, j. Clearly, the functions b^i_{jk} determine completely the Levi-Civita connection, and conversely. As $\nabla g = 0$, we have

$$b^i_{kj} = -b^i_{jk}, \quad (\text{in particular, } b^i_{jj} = 0) \tag{2}$$

for all i, j, k. We now put

$$b^1_{12} = \alpha,\ b^1_{13} = \beta,\ b^1_{23} = \gamma,\ b^2_{12} = \kappa,\ b^2_{13} = \mu,\ b^2_{23} = \nu,\ b^3_{12} = \sigma,\ b^3_{13} = \tau,\ b^3_{23} = \psi. \tag{3}$$

By (1)–(3) we get

$$\begin{aligned} &[e_1, e_2] = -\varepsilon\alpha\, e_1 - \kappa\, e_2 + (\varepsilon\gamma - \mu)\, e_3, \quad [e_1, e_3] = -\beta\, e_1 - (\gamma + \sigma)\, e_2 - \tau e_3, \\ &[e_2, e_3] = (\varepsilon\sigma - \mu)\, e_1 - \nu\, e_2 - \varepsilon\psi\, e_3. \end{aligned} \tag{4}$$

Conversely, the functions (b^i_{jk}) are determined by (4) via the *Koszul formula* [12].

A locally homogeneous Lorentzian three-manifold admits (locally) a pseudo-orthonormal basis $\{e_i\}$, such that (4) holds with *constant* connection functions $\alpha, \dots, \psi$. Starting from (4), we compute the curvature components with respect to $\{e_1\}$ and, by contraction, the Ricci components. We get

$$\rho_{11} = -\alpha^2 - \kappa^2 + \beta\nu - \gamma\mu + \sigma(\gamma - \mu) + \beta^2 - \tau^2 - \gamma\sigma + \alpha\psi + \mu(\gamma - \sigma), \quad (5)$$
$$\rho_{22} = -\alpha^2 - \kappa^2 + \beta\nu - \gamma\mu + \sigma(\gamma - \mu) + \nu^2 - \psi^2 - \kappa\tau + \mu\sigma + \gamma(\mu + \sigma), \quad (6)$$
$$\rho_{33} = -\beta^2 + \tau^2 + \gamma\sigma - \alpha\psi - \mu(\gamma - \sigma) - \nu^2 + \psi^2 + \kappa\tau - \mu\sigma - \gamma(\mu + \sigma), \quad (7)$$
$$\rho_{12} = \beta(\gamma + \sigma) + \nu(\gamma - \sigma) - \tau(\alpha + \psi), \quad (8)$$
$$\rho_{13} = -\alpha(\mu + \sigma) - \nu(\kappa - \tau) - \psi(\mu - \sigma), \quad (9)$$
$$\rho_{23} = \alpha(\beta - \nu) + \kappa(\gamma + \mu) - \tau(\gamma - \mu). \quad (10)$$

For the components of the covariant derivative of ρ with respect to $\{e_i\}$, we find

$$\nabla_i \rho_{jk} = -\sum_t \left(\varepsilon_j b^i_{jt}\rho_{tk} + \varepsilon_k B^i_{kt}\rho_{tj}\right). \quad (11)$$

Observe that the connection functions $\alpha, \dots, \psi$ are not all independent. In fact, since (M, g) is locally homogeneous, its scalar curvature $r = \operatorname{tr}\rho$ is constant. The well-known *divergence formula* $dr = 2\operatorname{div}\rho$ (see [12]) then implies $\sum_j \nabla_j \rho_{ij} = 0$, for all i, which, taking into account (11), gives some restrictions for the connection functions.

We end this section with the following classification result.

Theorem 1 ([1]) *A three-dimensional connected, simply connected complete homogeneous Lorentzian manifold (M, g) is either symmetric, or $M = G$ is a Lie group and g is left-invariant. Precisely, one of the following cases occurs:*

(I) If G is unimodular, then there exists a pseudo-orthonormal frame field $\{e_i\}$, with e_3 time-like, such that the Lie algebra of G is one of the following:

$$\mathfrak{g}_1: \ [e_1, e_2] = \alpha e_1 - \beta e_3, \ [e_1, e_3] = -\alpha e_1 - \beta e_2, \ [e_2, e_3] = \beta e_1 + \alpha e_2 + \alpha e_3, \ \alpha \neq 0. \quad (12)$$

If $\beta \neq 0$, then G is $\widetilde{SL}(2, \mathbb{R})$, while $G = E(1, 1)$ when $\beta = 0$.

$$\mathfrak{g}_2: \ [e_1, e_2] = -\gamma e_2 - \beta e_3, \ [e_1, e_3] = -\beta e_2 + \gamma e_3, \ [e_2, e_3] = \alpha e_1, \quad \gamma \neq 0. \quad (13)$$

In this case, $G = \widetilde{SL}(2, \mathbb{R})$ if $\alpha \neq 0$, while $G = E(1, 1)$ if $\alpha = 0$.

$$\mathfrak{g}_3: \quad [e_1, e_2] = -\gamma e_3, \quad [e_1, e_3] = -\beta e_2, \quad [e_2, e_3] = \alpha e_1. \quad (14)$$

The following Table 1 lists all the Lie groups G which admit a Lie algebra $\mathfrak{g}_3$, according to the different possibilities for α, β and γ:

Table 1 3D Lorentzian Lie groups with Lie algebra $\mathfrak{g}_3$

Lie group	(α, β, γ)	Lie group	(α, β, γ)
$\widetilde{SL}(2, \mathbb{R})$	$(+, +, +)$	$E(1, 1)$	$(+, -, 0)$
$\widetilde{SL}(2, \mathbb{R})$	$(+, -, -)$	$E(1, 1)$	$(+, 0, +)$
$SU(2)$	$(+, +, -)$	H_3	$(+, 0, 0)$
$\widetilde{E}(2)$	$(+, +, 0)$	H_3	$(0, 0, -)$
$\widetilde{E}(2)$	$(+, 0, -)$	$\mathbb{R}^3$	$(0, 0, 0)$

Table 2 3D Lorentzian Lie groups with Lie algebra $\mathfrak{g}_4$

Lie group ($\varepsilon = 1$)	α	β	Lie group ($\varepsilon = -1$)	α	β
$\widetilde{SL}(2, \mathbb{R})$	$\neq 0$	$\neq 1$	$\widetilde{SL}(2, \mathbb{R})$	$\neq 0$	$\neq -1$
$E(1, 1)$	0	$\neq 1$	$E(1, 1)$	0	$\neq -1$
$E(1, 1)$	<0	1	$E(1, 1)$	>0	-1
$\widetilde{E}(2)$	>0	1	$\widetilde{E}(2)$	<0	-1
H_3	0	1	H_3	0	-1

$$\mathfrak{g}_4 : \ [e_1, e_2] = -e_2 + (2\varepsilon - \beta)e_3, \ [e_1, e_3] = -\beta e_2 + e_3, \ [e_2, e_3] = \alpha e_1, \ \varepsilon = \pm 1. \tag{15}$$

Table 2 describes all Lie groups G admitting a Lie algebra $\mathfrak{g}_4$.

(II) If G is non-unimodular, there exists a pseudo-orthonormal frame field $\{e_i\}$, with e_3 time-like, such that $\alpha + \delta \neq 0$ and the Lie algebra of G is one of the following:

$$\mathfrak{g}_5 : [e_1, e_2] = 0, \ [e_1, e_3] = \alpha e_1 + \beta e_2, \ [e_2, e_3] = \gamma e_1 + \delta e_2, \ \ \alpha\gamma + \beta\delta = 0. \tag{16}$$

$$\mathfrak{g}_6 : \ [e_1, e_2] = \alpha e_2 + \beta e_3, \ [e_1, e_3] = \gamma e_2 + \delta e_3, \ [e_2, e_3] = 0, \ \ \alpha\gamma - \beta\delta = 0. \tag{17}$$

$$\begin{aligned} \mathfrak{g}_7 : -[e_1, e_2] &= [e_1, e_3] = \alpha e_1 + \beta e_2 + \beta e_3, \\ [e_2, e_3] &= \gamma e_1 + \delta e_2 + \delta e_3, \ \alpha\gamma = 0. \end{aligned} \tag{18}$$

3 The Basic System of Equations

We shall express Eqs. (5)–(10) via a system of PDE's, whose solutions give explicitly locally homogeneous Lorentzian metrics on $\mathbb{R}^3$ with the required curvature.

Fix a point $p \in M$ and consider a pseudo-orthonormal frame field $\{e_i\}$, satisfying (4) for some constants $\alpha, \ldots, \psi$. Choose a surface S through p transversal to the lines generated by e_3, a local coordinates system (w, x) on S and a neighborhood U_p of p, sufficiently small that each $q \in U_p$ is situated on exactly one line generated by e_3 and passing through one point $\bar{q} \in S$. Choose an orientation of S and define the coordinate function y in U_p as the oriented distance of q from S along the corresponding line, that is, $y(q) = \text{dist}(q, \pi(q))$, where $\pi : U_p \to S$ is the corresponding projection. We also define $w(q) = w(\pi(q)), x(q) = x(\pi(q))$. In this way, a local coordinate system (w, x, y) is introduced in U_p. Observe that $e_3 = \partial/\partial y$ and the coframe $\{\omega_1, \omega_2, \omega_3\}$ of $\{e_1, e_2, e_3\}$ must take the form

$$\omega^1 = Adw + Bdx, \quad \omega^2 = Cdw + Ddx, \quad \omega^3 = Gdw + Hdx + dy, \tag{19}$$

for some functions A, B, C, D, G, H. Next, we introduce the connection forms $\omega^i_j = \sum_k \varepsilon_j b^i_{jk} \omega^k$, which completely determine the Levi-Civita connection, because $\nabla_{e_i} e_j = \sum_k \omega^k_j(e_i) e_k$, for all i, j. Moreover, from (1) we easily get

$$\omega^i_j + \varepsilon_i \varepsilon_j \omega^j_i = 0 \tag{20}$$

for all i, j (in particular, $\omega^i_i = 0$ for all i). The *structure equations* for ω^i_j give

$$d\omega^i + \sum_j \omega^i_j \wedge \omega^j = 0, \tag{21}$$

for all indices i. The curvature forms Ω^i_j are completely determined by

$$-d\Omega^i_j = d\omega^i_j + \sum_k \omega^i_k \wedge \omega^k_j. \tag{22}$$

By the definition of the Ricci tensor and taking into account (20) and (4), we obtain that (22) is equivalent to

$$\begin{aligned}
d\omega^1_2 + \omega^1_3 \wedge \omega^3_2 &= -R_{1212}\, \omega^1 \wedge \omega^2 - \rho_{23}\, \omega^1 \wedge \omega^3 + \rho_{13}\, \omega^2 \wedge \omega^3, \\
d\omega^1_3 + \omega^1_2 \wedge \omega^2_3 &= \rho_{23}\, \omega^1 \wedge \omega^2 + R_{1313}\, \omega^1 \wedge \omega^3 - \rho_{12}\, \omega^2 \wedge \omega^3, \\
d\omega^2_3 + \omega^2_1 \wedge \omega^1_3 &= -\rho_{13}\, \omega^1 \wedge \omega^2 - \rho_{12}\, \omega^1 \wedge \omega^3 + R_{2323}\, \omega^2 \wedge \omega^3.
\end{aligned} \tag{23}$$

We then use (19) in (21). Also taking into account (3) and the divergence formula, we obtain that (21) is equivalent to the following system of nine PDE's:

$$\begin{array}{ll}
A'_y = \beta A + (\mu + \sigma) C, & B'_y = \beta B + (\mu + \sigma) D, \\
C'_y = (\gamma - \sigma) A + \nu C, & D'_y = (\gamma - \sigma) B + \nu D, \\
G'_y = -\tau A - \psi C, & H'_y = -\tau B - \psi D, \\
B'_w - A'_x = \alpha \mathscr{D} - \beta \mathscr{E} - (\mu + \sigma) \mathscr{F}, & D'_w - C'_x = \kappa \mathscr{D} - (\gamma - \sigma) \mathscr{E} - \nu \mathscr{F}, \\
H'_w - G'_x = -(\gamma - \mu) \mathscr{D} + \tau \mathscr{E} + \psi \mathscr{F}, &
\end{array} \tag{24}$$

where $\mathscr{D}, \mathscr{E}, \mathscr{F}$ are auxiliary functions, defined by

$$\mathscr{D} = AD - BC, \qquad \mathscr{E} = AH - BG, \qquad \mathscr{F} = CH - DG. \tag{25}$$

Observe that, because of (19), $\mathscr{D} = AD - BC \neq 0$ is a necessary and sufficient condition for linear independence of the ω^i. Starting from the connection functions b^i_{jk} of (M, g), by (24) we determine the functions $A, \dots, H$ and so, explicit Lorentzian metrics on $\mathbb{R}^3$, with the same Levi-Civita connection of (M, g). Conversely, if $A, \dots, H$ are known, then by (24) we can determine b^i_{jk}.

We now express the curvature conditions (23) using (19). Taking into account that the connection functions are constant, one can easily prove that (23) is equivalent to the following system of algebraic equations:

$$\begin{array}{l}
(U_3 + R_{1212})\mathscr{D} + (V_3 + \rho_{23})\mathscr{E} + (W_3 - \rho_{13})\mathscr{F} = 0,\\
(U_2 - \rho_{23})\mathscr{D} + (V_2 - R_{1313})\mathscr{E} + (W_2 + \rho_{12})\mathscr{F} = 0,\\
(U_1 + \rho_{13})\mathscr{D} + (V_1 + \rho_{12})\mathscr{E} + (W_1 - R_{2323})\mathscr{F} = 0,\\
(V_3 + \rho_{23})A + (W_3 - \rho_{13})C = 0, \quad (V_3 + \rho_{23})B + (W_3 - \rho_{13})D = 0,\\
(V_2 - R_{1313})A + (W_2 + \rho_{12})C = 0, \quad (V_2 - R_{1313})B + (W_2 + \rho_{12})D = 0,\\
(V_1 + \rho_{12})A + (W_1 - R_{2323})C = 0, \quad (V_1 + \rho_{12})B + (W_1 - R_{2323})D = 0,
\end{array}$$

where we put

$$\begin{aligned}
U_1 &= \alpha(\gamma + \mu) - \kappa(\beta - \nu) - \psi(\gamma - \mu),\\
V_1 &= -\beta(\gamma + \sigma) - \nu(\gamma - \sigma) + \tau(\alpha + \psi),\\
W_1 &= -\nu^2 + \psi^2 + \kappa\tau - \mu\sigma - \gamma(\mu + \sigma),\\
U_2 &= \alpha(\beta - \nu) + \kappa(\gamma + \mu) - \tau(\gamma - \mu),\\
V_2 &= -\beta^2 + \tau^2 - \alpha\psi + \gamma\sigma - \mu(\gamma - \sigma),\\
W_2 &= -\beta(\mu + \sigma) - \nu(\mu - \sigma) - \psi(\kappa + \tau),\\
U_3 &= \alpha^2 + \kappa^2 - \beta\nu + \gamma\mu - \sigma(\gamma - \mu),\\
V_3 &= -\beta(\alpha + \psi) - \kappa(\gamma - \sigma) + \tau(\gamma + \sigma),\\
W_3 &= -\alpha(\mu + \sigma) - \nu(\kappa - \tau) - \psi(\mu - \sigma).
\end{aligned} \tag{26}$$

Comparing (8)–(10) with (26), we easily get $V_1 + \rho_{12} = 0$, $U_2 - \rho_{23} = 0$ and $W_3 - \rho_{13} = 0$. Hence, Eq. (26) reduce to

$$\begin{array}{l}
(U_3 + R_{1212})\mathscr{D} + (V_3 + \rho_{23})\mathscr{E} = 0, \quad (V_3 + \rho_{23})A = 0,\\
(V_2 - R_{1313})\mathscr{E} + (W_2 + \rho_{12})\mathscr{F} = 0, \quad (V_3 + \rho_{23})B = 0,\\
(V_2 - R_{1313})A + (W_2 + \rho_{12})C = 0, \quad (V_2 - R_{1313})B + (W_2 + \rho_{12})D = 0,\\
(U_1 + \rho_{13})\mathscr{D} + (W_1 - R_{2323})\mathscr{F} = 0, \ (W_1 - R_{2323})C = 0,\\
(W_1 - R_{2323})D = 0.
\end{array} \tag{27}$$

In this way, we have proved the following result.

Theorem 2 *Given a locally homogeneous Lorentzian three-manifold* (M, g)*, having* $\mathscr{R} = (\rho_{ij})$ *as the matrix of Ricci components with respect to a suitable* *pseudo-orthonormal frame* $\{e_i\}$, *let* A, B, C, D, G, H *be smooth functions on* (w, x, y), *satisfying the systems (24) and (27). Then, (19) determines a locally homogeneous Lorentzian metric* $\bar{g}$ *on* $\mathbb{R}^3$, *locally isometric to* (M, g) *(in particular, having the same curvature).*

4 Explicit Lorentzian Metrics in $\mathbb{R}^3$ with Prescribed Curvature

For each of the homogeneous models described by (12)–(18), we can now solve systems (24) and (27), providing explicit Lorentzian metrics on $\mathbb{R}^3$ which have exactly the Ricci tensor of the corresponding model. Curvature equations are remarkably simpler when the Ricci tensor is diagonal. This special case has been studied in [4]. Hence, we focus here on all the remaining cases, which do not have any correspondence with the Riemannian case. The Ricci tensor of all 3*D* Lie groups equipped with a left-invariant Lorentzian metric was calculated in [2] and can be easily obtained by direct calculation starting from (12)–(18). According to the results of [2], non-diagonal cases occur for the Lie algebras $\mathfrak{g}_1$, $\mathfrak{g}_2$, $\mathfrak{g}_4$ and $\mathfrak{g}_7$.

($\mathfrak{g}_1$) Comparing (12) with (4), we find that the connection functions of a locally homogeneous Lorentzian three-manifold described by (12) are given by

$$\alpha = \beta = -\nu = \psi = -a, \quad \gamma = -\mu = -\sigma = -\tfrac{b}{2}, \quad \kappa = \tau = 0, \tag{28}$$

where $a \neq 0$ and b are constant. Straightforward calculations (see also [1]) show that the Ricci tensor at any point is given by

$$\mathscr{R}_1 = \begin{pmatrix} -\frac{b^2}{2} & -ab & ab \\ -ab & -2a^2 - \frac{b^2}{2} & 2a^2 \\ ab & 2a^2 & \frac{b^2}{2} - 2a^2 \end{pmatrix}. \tag{29}$$

On the other hand, because of (28), Eq. (26) reduce to

$$\begin{array}{lll} U_1 = -ab, & U_2 = 2a^2, & U_3 = 2a^2 + \frac{b^2}{4}, \\ V_1 = ab, & V_2 = -2a^2 + \frac{b^2}{4}, & V_3 = ab, \\ W_1 = \frac{b^2}{4}, & W_2 = ab, & W_3 = ab. \end{array} \tag{30}$$

By (29) and (30) it follows at once that *all Eqs.* (27) *reduce to identities*, that is, under the assumption (28), the curvature conditions (27) are identically satisfied.

We now turn our attention to the connection equations (24). Again by (28), we obtain that (24) reduces to

$$
\begin{array}{lll}
A'_y = -aA + bC, & B'_y = -aB + bD, & C'_y = -bA + aC, \\
D'_y = -bB + aD, & G'_y = aC, & H'_y = aD, \\
B'_w - A'_x = -a\mathscr{D} + a\mathscr{E} - b\mathscr{F}, & D'_w - C'_x = b\mathscr{E} - a\mathscr{F}, & H'_w - G'_x = b\mathscr{D} - a\mathscr{F}.
\end{array}
\tag{31}
$$

One can now find explicit solutions of the system (31). Different kinds of solutions are obtained according to the different possibilities for the sign of $a^2 - b^2$. Some explicit solutions of (31) are resumed in the following

Theorem 3 *Let $a \neq 0$ and b be two real constants and $\mathscr{R}_1$ any symmetric real matrix described by (29). Then, (19) determines a family of (locally isometric) locally homogeneous Lorentzian metrics on $\mathbb{R}^3[w,x,y]$ having $\mathscr{R}_1$ as the Ricci tensor at any point, where the functions A, B, C, D, G, H are the following:*
(i) When $b \neq 0$ and $a^2 - b^2 > 0$, we put $\eta = \sqrt{a^2 - b^2}$. Then

$$
\begin{array}{ll}
A = f\cosh(\eta y), & B = \theta \sinh(\eta y), \\
C = \frac{1}{b} f\,(a\cosh(\eta y) + \eta \sinh(\eta y)), & D = \frac{1}{b}\theta\,(\eta\cosh(\eta y) + a\sinh(\eta y)), \\
G = \frac{a}{b\eta} f\,(\eta\cosh(\eta y) + a\sinh(\eta y)) - \frac{1}{\theta\eta} f'_x, & \\
H = \frac{a}{b\eta}\theta\,(a\cosh(\eta y) + \eta\sinh(\eta y)), &
\end{array}
$$

for a real constant $\theta \neq 0$ and $f(w,x) = a_1(w)\cos(b\theta x) + a_2(w)\sin(b\theta x)$, where a_1, a_2 are two arbitrary one-variable functions. Corresponding solutions are found in [3] in the cases $a^2 = b^2$ and $a^2 - b^2 < 0$. In all the cases, the corresponding Lorentzian metric is defined in the open subset of $\mathbb{R}^3$ where $f \neq 0$.
(ii) When $b = 0$:

$$
A = a_0(w)e^{-ay}, \quad B = b_0(x)e^{-ay}, \quad C = G = c_0(w)e^{ay}, \quad D = H = d_0(x)e^{ay},
$$

where a_0, b_0, c_0, d_0 are arbitrary one-variable functions. The corresponding Lorentzian metric is defined in the open subset of $\mathbb{R}^3$ where $a_0(w)d_0(x) - b_0(x)c_0(w) \neq 0$.

($\mathfrak{g}_2$) The remaining cases can be treated similarly to the case $\mathfrak{g}_1$ above. So, for any of them, we shall only report the Ricci components, the equations for the connection functions and some explicit solutions. In the case of $\mathfrak{g}_2$, we have

$$
\mathscr{R}_2 = \begin{pmatrix} -\frac{a^2}{2} - 2c^2 & 0 & 0 \\ 0 & \frac{a^2}{2} - ab & c(a-2b) \\ 0 & c(a-2b) & -\frac{a^2}{2} + ab \end{pmatrix},
\tag{32}
$$

for three real constants a, b, c, and

$$
\begin{array}{lll}
A'_y = aC, & B'_y = aD, & C'_y = -bA, \\
D'_y = -bB, & G'_y = cA, & H'_y = cB, \\
B'_w - A'_x = -a\mathscr{F}, & D'_w - C'_x = c\mathscr{D} + b\mathscr{E}, & H'_w - G'_x = b\mathscr{D} - c\mathscr{E}.
\end{array}
\tag{33}
$$

We present some solutions of (33) in the following

Theorem 4 *Given three real constants a, b, c and any symmetric real matrix $\mathscr{R}_2$ described by (32). Then, (19) gives a family of (locally isometric) locally homogeneous Lorentzian metrics on $\mathbb{R}^3[w, x, y]$ having $\mathscr{R}_2$ as the Ricci tensor at any point, where the functions A, B, C, D, G, H are the following:*

If $-ab < 0$, we put $\eta = \sqrt{ab}$. Then

$$\begin{array}{lll} A = f\cos(\eta y), & B = \theta\sin(\eta y), & C = -\frac{\eta}{a} f\sin(\eta y), \\ D = \frac{\eta}{a}\theta\cos(\eta y), & G = \frac{c}{\eta} f\sin(\eta y) - \frac{1}{\theta\eta} f'_x, & H = -\frac{c}{\eta}\theta\cos(\eta y), \end{array}$$

where $f(w, x) = a_1(w)\cosh(\sqrt{\theta^2(b^2 + c^2)}x) + a_2(w)\sinh(\sqrt{\theta^2(b^2 + c^2)}\theta x), \theta \neq 0$ is a real constant and a_1, a_2 are two arbitrary one-variable functions. The Lorentzian metric is defined on the open subset of $\mathbb{R}^3$ where $f \neq 0$. Corresponding solutions were found in [3] in the cases $ab < 0$, $a = 0$, $b = 0$.

($\mathfrak{g}_4$) For a locally homogeneous Lorentzian three-manifold described by (15), the Ricci components are given by

$$\mathscr{R}_4 = \begin{pmatrix} -\frac{a^2}{2} & 0 & 0 \\ 0 & \frac{a^2}{2} + 2\varepsilon(a - b) - ab + 2 & a + 2(\varepsilon - b) \\ 0 & a + 2(\varepsilon - b) & -\frac{a^2}{2} + ab + 2 - 2\varepsilon b \end{pmatrix}, \tag{34}$$

for two real constants a, b, and connection equations (24) become

$$\begin{array}{lll} A'_y = aC, & B'_y = aD & C'_y = -bA, \\ D'_y = -bB, & G'_y = A, & H'_y = B, \\ B'_w - A'_x = -a\mathscr{F}, & D'_w - C'_x = \mathscr{D} + b\mathscr{E}, & H'_w - G'_x = (b - 2\varepsilon)\mathscr{D} - \mathscr{E}. \end{array} \tag{35}$$

Some explicit solutions of (35) are given in the following

Theorem 5 *Given two real constants a, b and any symmetric real matrix $\mathscr{R}_4$ as in (34). Then, (19) describes a family of (locally isometric) locally homogeneous Lorentzian metrics on $\mathbb{R}^3[w, x, y]$ whose Ricci tensor at any point is $\mathscr{R}_4$, where the functions A, B, C, D, G, H are the following:*

If $ab < 0$, we put $\eta = \sqrt{-ab}$. Then,

$$\begin{array}{lll} A = f\cosh(\eta y), & B = \theta\sinh(\eta y), & C = \frac{\eta}{a} f\sinh(\eta y), \\ D = \frac{\eta}{a}\theta\cosh(\eta y), & G = \frac{1}{\eta} f\sinh(\eta y) - \frac{1}{\theta\eta} f'_x, & H = \frac{1}{\eta}\theta\cosh(\eta y), \end{array}$$

where

$$f(w, x) = \begin{cases} a_1(w)\cos(|\theta(b + \varepsilon)|x) + a_2(w)\sin(|\theta(b + \varepsilon)|x) & \text{if } b \neq -\varepsilon, \\ a_1(w)x + a_2(w) \quad \text{if } b = -\varepsilon, \end{cases}$$

for a real constant $\theta \neq 0$ and two arbitrary one-variable functions a_1, a_2. The Lorentzian metric is defined in the open subset of $\mathbb{R}^3$ where $f \neq 0$. Corresponding solutions were found in [3] in the cases $ab > 0$, $a = 0$, $b = 0$.

($\mathfrak{g}_7$) Consider a locally homogeneous Lorentzian three-manifold locally described by (18). Then, the Ricci components are given by

$$\mathscr{R}_7 = \begin{pmatrix} -\frac{c^2}{2} & 0 & 0 \\ 0 & ad - a^2 - bc + \frac{c^2}{2} & a^2 - ad + bc \\ 0 & a^2 - ad + bc & ad - a^2 - bc - \frac{c^2}{2} \end{pmatrix}, \tag{36}$$

where a, b, c, d are four real constants satisfying $ac = 0$.

If $c = 0$, then, (24) reduces to

$$\begin{array}{ll} A'_y = aA, & B'_y = aB, \\ C'_y = G'_y = bA + dC, & D'_y = H'_y = bB + dD, \\ B'_w - A'_x = a\mathscr{D} - a\mathscr{E}, & D'_w - C'_x = H'_w - G'_x = b\mathscr{D} - b\mathscr{E} - d\mathscr{F}, \end{array} \tag{37}$$

while if $c \neq 0$, then $a = 0$ and the system (24) reduces to

$$\begin{array}{ll} A'_y = cC, & B'_y = cD, \\ C'_y = G'_y = bA + dC, & D'_y = H'_y = bB + dD, \\ B'_w - A'_x = -c\mathscr{F}, & D'_w - C'_x = H'_w - G'_x = b\mathscr{D} - b\mathscr{E} - d\mathscr{F}. \end{array} \tag{38}$$

Some solutions of (37) and (38) are given in the following

Theorem 6 *Given three real constants a, b, d and any symmetric real matrix $\mathscr{R}_7$ described by (36). Then, (19) gives a family of (locally isometric) locally homogeneous Lorentzian metrics on $\mathbb{R}^3[w, x, y]$ having $\mathscr{R}_7$ as the Ricci tensor at any point, where the functions A, B, C, D, G, H are the following:*

(I) When $c = 0$:

$$\begin{array}{ll} A = a_0(w)e^{ay}, & B = b_0(x)e^{ay}, \\ C = G = e^{dy}(c_0(w) + \frac{b}{a-d}a_0(w)e^{(a-d)y}), & D = H = e^{dy}(d_0(x) + \frac{b}{a-d}b_0(x)e^{(a-d)y}), \end{array}$$

where a_0, b_0, c_0, d_0 are arbitrary one-variable functions. The Lorentzian metric is defined in the open subset of $\mathbb{R}^3$ where $a_0(w)d_0(x) - b_0(x)c_0(w) \neq 0$.

(II) When $a = 0 \neq c$: if $\Delta = d^2 + 4bc > 0$, let $\lambda_1 \neq \lambda_2$ be the solutions of $\lambda^2 - d\lambda - bc = 0$. Then,

$$\begin{array}{ll} A = k_1(w)e^{\lambda_1 y} + k_2(w)e^{\lambda_2 y}, & B = h_1(x)e^{\lambda_1 y} + h_2(x)e^{\lambda_2 y}, \\ C = G = \frac{1}{c}(k_1(w)\lambda_1 e^{\lambda_1 y} + k_2(w)\lambda_2 e^{\lambda_2 y}), & D = H = \frac{1}{c}(h_1(x)\lambda_1 e^{\lambda_1 y} + h_2(x)\lambda_2 e^{\lambda_2 y}), \end{array}$$

where k_1, k_2, h_1, h_2 are four arbitrary one-variable functions, and the Lorentzian metric is defined on the open subset of $\mathbb{R}^3$ where $k_1(w)h_2(x) - k_2(w)h_1(x) \neq 0$. Corresponding solutions exist when $\Delta = 0$ and when $\Delta < 0$ (see [3]).

Acknowledgments Partially supported by the Ministry of Economy and Competitiveness, Spain, project MTM2011-22528.

References

1. Calvaruso, G.: Homogeneous structures on three-dimensional Lorentzian manifolds. J. Geom. Phys. **57**, 1279–1291 (2007); Addendum: J. Geom. Phys. **58**, 291–292 (2008)
2. Calvaruso, G.: Einstein-like metrics on three-dimensional homogeneous Lorentzian manifolds. Geometriae Dedicata **127**, 99–119 (2007)
3. Calvaruso, G.: Three-dimensional homogeneous Lorentzian metrics with prescribed Ricci tensor. J. Math. Phys. **48**, 123518, 17 p. (2007)
4. Calvaruso, G.: Pseudo-Riemannian 3-manifolds with prescribed distinct constant Ricci eigenvalues. Differ. Geom. Appl. **26**, 419–433 (2008)
5. Calvaruso, G.: Curvature homogeneous Lorentzian three-manifolds. Ann. Glob. Anal. Geom. **36**, 1–17 (2009)
6. DeTurck, D.M.: The equation of prescribed Ricci curvature. Bull. Am. Math. Soc. **3**, 701–704 (1980)
7. DeTurck, D.M.: Existence of metrics with prescribed Ricci curvature: local theory. Invent. Math. **65**, 179–207 (1981)
8. DeTurck, D.M.: The Cauchy problem for Lorentz metrics with prescribed Ricci curvature. Compos. Math. **48**, 327–349 (1983)
9. DeTurck, D.M., Goldschmidt, H., Talvacchia, J.: Connections with prescribed curvature and Yang-Mills currents: the semi-simple case. Ann. Sci. Ecole Norm. S. **24**, 57–112 (1991)
10. DeTurck, D.M., Goldschmidt, H., Talvacchia, J.: Local existence of connections with prescribed curvature. In: Differential Geometry, Global Analysis, and Topology. Halifax, NS (1990), pp. 13-25. In: CMS Conf. Proc., vol. 12, Amer. Math. Soc., Providence, RI (1991)
11. Muñoz-Masqué, J., Pozo Coronado, L.M., Sánchez Rodríguez, I. J.: The prescribed curvature problem in dimension four. Math. Pure Appl. **92**, 599–612 (2009)
12. O'Neill, B.: Semi-Riemannian Geometry. Academic Press, New York (1983)
13. Sekigawa, K.: On some three-dimensional curvature homogeneous spaces. Tensor **31**, 87–97 (1977)
14. Tsarev, S.P.: Which 2-forms are locally curvature forms? Funct. Anal. Appl. **16**, 235–237 (1982)

Euler–Poincaré Reduction by a Subgroup of Symmetries as an Optimal Control Problem

Marco Castrillón López and Pedro L. García

Dedicated to our friend and colleague Jaime Muñoz Masqué on the occasion of his 65th birthday

Abstract Given a Lagrangian density Ldt defined in the 1-jet bundle J^1P of a principal G-bundle $P \to \mathbb{R}$, invariant with respect to the action of a closed subgroup $H \subset G$, its Euler–Poincaré reduction in $(J^1P)/H = C(P) \times_{\mathbb{R}} (P/H)$ ($C(P)$: the bundle of connections, P/H: the bundle of H-structures) induces an optimal control problem. The control variables of this problem are connections σ, the dynamical variables $\bar{s}$ are H-structures, the Lagrangian density $l(t, \sigma, \bar{s})dt$ is the reduction of Ldt and the dynamical equations are $\nabla^\sigma \bar{s} = 0$. We prove that the solution of this problem are solutions of the original reduction problem. We study the Hamilton–Cartan–Pontryagin formulation of the problem under an appropriate regularity condition. Finally, the theory is illustrated with the example of the heavy top, for which the symplectic structure of the set of solutions with zero vertical component of the angular momentum is also provided.

Keywords Mechanics · Optimal control · Reduction · Symmetries · Variational calculus

M. Castrillón López (✉)
ICMAT(CSIC-UAM-UC3M-UCM), Departamento de Geometría y Topología,
Facultad de CC. Matemáticas, Universidad Complutense de Madrid, 28040 Madrid, Spain
e-mail: mcastri@mat.ucm.es

P.L. García
Real Academia de Ciencias Exactas Físicas y Naturales and IUFFyM,
Universidad de Salamanca, 37008 Salamanca, Spain
e-mail: pgarcia@usal.es

M. Castrillón López et al. (eds.), *Geometry, Algebra and Applications: From Mechanics to Cryptography*, Springer Proceedings in Mathematics & Statistics 161, DOI 10.1007/978-3-319-32085-4_5

1 Introduction

In [3], the authors proposed a new formulation of the Euler–Poincaré reduction scheme in principal bundles by a subgroup of the structure group by means of a canonical reduction morphism. More precisely, given a Lagrangian density $L\mathbf{v}$ defined in the fiber bundle J^1P of 1-jets of local section of a principal G-bundle $\pi : P \to M$, invariant under the action of a closed subgroup $H \subset G$, the corresponding variational problem projects to $(J^1P)/H$ which can be identified to

$$\begin{aligned}(J^1P)/H &\widetilde{\longrightarrow} C(P) \times_M (P/H)\\ [j^1_x s]_H &\mapsto ([j^1_x s]_G, [s(x)]_H)\end{aligned}$$

where $(J^1P)/G = C(P) \to M$ is the bundle of connections of P. This reduced problem is thus defined on connections $\sigma \in \Gamma(M, C(P))$ and H-structures $\bar{s} \in \Gamma(M; P/H)$ by a Lagrangian density $l\mathbf{v}$, the projection of $L\mathbf{v}$, the constraints $\mathrm{Curv}\sigma = 0$ and $\nabla^\sigma \bar{s} = 0$, and the representation in connections and H-structures of infinitesimal gauge transformations $\eta \in \Gamma(M, \tilde{\mathfrak{g}})$ as the set of admissible infinitesimal variations ($\tilde{\mathfrak{g}} \to M$ being the adjoint bundle).

Due to the gauge functoriality of the curvature and the covariant derivative, given an admissible section $(\sigma, \bar{s}) \in \Gamma(M, C(P) \times_M (P/H))$, i.e., a section satisfying $\mathrm{Curv}\sigma = 0$ and $\nabla^\sigma \bar{s} = 0$, the 1-jet extension $j^1(\delta\sigma, \delta\bar{s})$ of an admissible infinitesimal variations $(\delta\sigma, \delta\bar{s})$ are tangent to the submanifold

$$S = \{j^1_x(\sigma, \bar{s}) : \mathrm{Curv}\sigma = 0, \nabla^\sigma \bar{s} = 0\} \subset J^1(C(P) \times_M (P/H))$$

along $j^1(\sigma, \bar{s})$. Therefore, we obtain a subspace of the space of admissible infinitesimal variations along an admissible section of the Lagrange problem defined in $J^1(C(P) \times_M (P/H))$ by the reduced Lagrangian $l\mathbf{v}$ and the constraint submanifold S (see [5, 6] for a recent geometric version of the Lagrange problem). In other words, we can canonically associate to the Euler–Poincaré reduction by a subgroup of symmetries, a Lagrange problem the critical sections of which define a subset of the set of solutions of the Euler–Poincaré equations of the original variational problem. We think that the study of this situation is of interest in the framework of the Euler–Poincaré in fiber bundles.

In [2] we first tackled this study in the case $H = G$, where the reduction morphism is $J^1P \to (J^1P)/G = C(P)$ and no H-structures occur. The reduced problem is defined on connections $\sigma \in \Gamma(M, C(P))$ with the constraint $\mathrm{Curv}\sigma = 0$. In the work we now present, we study the case for arbitrary H in Mechanics, that is, when $M = \mathbb{R}$. Note that, in this situation, the condition $\mathrm{Curv}\sigma = 0$ is trivially satisfied and the only constraint is $\nabla^\sigma \bar{s} = 0$. In particular, from the local expression of $\nabla^\sigma \bar{s}$, the problem of Lagrange can be seen as an optimal control problem where the dynamic variable is the H structure $\bar{s}$ and the control variable is the connection σ (see [1, 5] for the notions on optimal control problems). We also study the regularity and the Hamilton–Cartan–Pontryagin formalism of the problem, with the conviction that these results will shed light to the general case with arbitrary manifold M.

The structure of the work is as follows. In Sect. 2, we give the local expressions of some basic operators appearing in the Euler–Poincaré reduction framework that will be useful in the following. In Sect. 3, we state the problem of Lagrange associated to an Euler–Poincaré reduction by a subgroup of symmetries, putting emphasis on its nature of an optimal control problem, with dynamical variable $\bar{s}$ and control σ. In Sect. 4 we compare the solution of the Lagrange problem obtained with Lagrange multipliers with the solutions of the Euler–Poincaré equations of the initial reduced problem. Section 5 gives the Hamilton–Cartan–Pontryagin formulation of the problem. Finally, Sect. 6 applies the results to the case of the heavy top, an example where in addition, we describe the symplectic structure of the set of solutions in the zero level set of the vertical component of the angular momentum.

2 Some Local Formulas

Let $\pi : P \to \mathbb{R}$ be a principal bundle with structure group a Lie group G with Lie algebra $\mathfrak{g}$. Let $H \subset G$ be a closed subgroup with Lie algebra $\mathfrak{h}$, and let $\pi_H : P \to P/H$ be the H-principal bundle over P/H, which in addition can be identified to the associated bundle $P \times_G (G/H) \to \mathbb{R}$ with respect of the natural left action of G on G/H. Let $V(P/H) \subset T(P/H)$ be the bundle of vertical vector fields tangent to the fibers of $P/H \to \mathbb{R}$. Following [3] (Sects. 2 and 3), given an infinitesimal gauge transformation $\eta \in \Gamma(\mathbb{R}, \tilde{\mathfrak{g}})$, understood as a G-invariant vertical vector field in P, we denote by $\eta_{P/H} \in \Gamma(P/H, V(P/H))$ its projection by π_H. Given a section $(\sigma, \bar{s}) \in \Gamma(\mathbb{R}, C(P) \times_{\mathbb{R}} (P/H))$, the infinitesimal transformations $\delta\sigma = \mathrm{P}_\sigma(\eta)$ and $\delta\bar{s} = \mathrm{P}_{\bar{s}}(\eta)$ induced along $(\sigma, \bar{s})$ by the infinitesimal gauge transformation $\eta \in \Gamma(\mathbb{R}, \tilde{\mathfrak{g}})$ reads as

$$\delta\sigma = \mathrm{P}_\sigma(\eta) = \nabla^\sigma \eta \in \Gamma(\mathbb{R}, \sigma^* V(C(P))), \quad \delta\bar{s} = \mathrm{P}_{\bar{s}}(\eta) = \eta_{P/H} \in \Gamma(\mathbb{R}, \bar{s}^* V(P/H)).$$

We note that the operator $\mathrm{P}_{\bar{s}}$ is surjective and its kernel is

$$\ker \mathrm{P}_{\bar{s}} = \{\eta \in \Gamma(\mathbb{R}, \tilde{\mathfrak{g}}) : \eta(x) = [u, B]_G \text{ with } [u]_H = \bar{s}(x) \text{ and } B \in \mathfrak{h}\}.$$

In addition, we have the isomorphism

$$\begin{aligned} \bar{s}^*\tilde{\mathfrak{h}} &\xrightarrow{\sim} \ker \mathrm{P}_{\bar{s}} \\ [u, B]_H &\mapsto [u, B]_G \end{aligned}$$

where $\tilde{\mathfrak{h}} \to P/H$ is the adjoint fiber bundle of the principal H-bundle $P \to P/H$.

Let $P = \mathbb{R} \times G \to \mathbb{R}$ be a trivialization of P. If $\{B_1, \ldots, B_m\}$, $m = \dim G$, is a basis of $\mathfrak{g}$, then $\{\tilde{B}_1, \ldots, \tilde{B}_m\}$ is a basis of the $C^\infty(\mathbb{R})$-module of section of $\tilde{\mathfrak{g}} \to M$, where $\tilde{B}_\alpha$ is the infinitesimal generator of the flow $((t, g), \varepsilon) \mapsto (t, \exp(\varepsilon B_\alpha)g)$ in P.

In addition, every G-invariant vector field in P can be written as $D = f\partial/\partial t + g^\alpha \tilde{B}_\alpha$, for $f, g^\alpha \in C^\infty(\mathbb{R})$. In particular, we can introduce coordinates in $C(P) \to \mathbb{R}$ as

$$\gamma_t\left(\frac{\partial}{\partial t}\right) = \frac{\partial}{\partial t} + A^\alpha(\gamma_t)\tilde{B}_\alpha, \qquad \gamma_t \in C(P),$$

where we understand γ_t as the horizontal lift $\gamma_t : T_t\mathbb{R} \to TP$ of the connection γ_t in the point $t \in \mathbb{R}$. Given a connection $\sigma \in \Gamma(\mathbb{R}, C(P))$, the connection 1-form along the trivial section of $P = \mathbb{R} \times G$ is

$$\omega_\sigma = -(A^\alpha \circ \sigma)dt \otimes B_\alpha,$$

from where we deduce (see, for instance, [7], Sect. 5.6) that for all $\eta = \eta^\alpha \tilde{B}_\alpha \in \Gamma(\mathbb{R}, \tilde{\mathfrak{g}})$

$$P_\sigma(\eta) = \nabla^\sigma \eta = \left(\frac{d\eta^\alpha}{dt}\tilde{B}_\alpha - (A^\alpha \circ \sigma)\eta^\beta[\tilde{B}_\beta, \tilde{B}_\alpha]\right) \otimes dt. \tag{1}$$

On the other hand, let $P/H = \mathbb{R} \times (G/H) \to \mathbb{R}$ be the trivialization induced by $P = \mathbb{R} \times G \to \mathbb{R}$ in P/H. If (y^i), $1 \leq i \leq r$, are coordinates in G/H, then the vector fields $(\tilde{B}_\alpha)_{P/H}$ in P/H can be expressed as

$$(\tilde{B}_\alpha)_{P/H} = \Psi^j_\alpha \frac{\partial}{\partial y^j}, \qquad \Psi^j_\alpha \in C^\infty(G/H). \tag{2}$$

In this situations for every $\bar{s} \in \Gamma(\mathbb{R}, P/H)$ and every $\eta = \eta^\alpha \tilde{B}_\alpha \in \Gamma(\mathbb{R}, \tilde{\mathfrak{g}})$, we have

$$\mathrm{P}_{\bar{s}}(\eta) = (\eta_{P/H})_{\bar{s}} = \eta^\alpha(\Psi^j_\alpha \circ \bar{s})\left(\frac{\partial}{\partial y^j}\right)_{\bar{s}}. \tag{3}$$

Finally, from the definition of covariant derivative of a section $\bar{s}$ with respect to a connection σ (see [3]), we have

$$\begin{aligned}
\nabla^\sigma_{\partial/\partial t}\bar{s} &= d\bar{s}\left(\frac{\partial}{\partial t}\right) - \left(\frac{\partial}{\partial t} + (A^\alpha \circ \sigma)\Psi^j_\alpha \frac{\partial}{\partial y^j}\right)_{\bar{s}} \\
&= \frac{\partial}{\partial t} + \frac{d(y^j \circ \bar{s})}{dt}\left(\frac{\partial}{\partial y^j}\right)_{\bar{s}} - \frac{\partial}{\partial t} - (A^\alpha \circ \sigma)\left(\Psi^j_\alpha \circ \bar{s}\right)\left(\frac{\partial}{\partial y^j}\right)_{\bar{s}} \\
&= \left(\frac{d(y^j \circ \bar{s})}{dt} - (A^\alpha \circ \sigma)\Psi^j_\alpha\right)\left(\frac{\partial}{\partial y^j}\right)_{\bar{s}},
\end{aligned}$$

so that

$$\nabla^\sigma \bar{s} = \left(\frac{d(y^j \circ \bar{s})}{dt} - (A^\alpha \circ \sigma)\left(\Psi^j_\alpha \circ \bar{s}\right)\right) dt \otimes \left(\frac{\partial}{\partial y^j}\right)_{\bar{s}}. \tag{4}$$

3 The Optimal Control Problem

We begin with the Euler–Poincaré reduction in a principal G-bundle $\pi : P \to \mathbb{R}$ by a closed subgroup $H \subset G$. We then consider the Lagrange problem defined in $J^1(C(P) \times_{\mathbb{R}} (P/H))$ by the reduced Lagrangian density $l dt$ and

$$S = \{j^1_t(\sigma, \bar{s}) : (\nabla^\sigma \bar{s})(t) = 0\} \subset J^1(C(P) \times_{\mathbb{R}} (P/H))$$

as constraint submanifold. From Eq. (4), we can locally characterize S as the set where the functions

$$\Phi^j = \dot{y}^j - A^\alpha \Psi^j_\alpha(y^1, \dots, y^r) = 0, \quad 1 \le j \le r \tag{5}$$

vanish, where $(t, A^\alpha, y^j, \dot{A}^\alpha, \dot{y}^j)$ are coordinates in $J^1(C(P) \times_{\mathbb{R}} (P/H))$ induced by the coordinates (t, A^α, y^j) of $C(P) \times_{\mathbb{R}} (P/H)$. Equation (5) suggest that the problem of Lagrange can be seen as an optimal control problem where the dynamic variable is $\bar{s} \in \Gamma(\mathbb{R}, P/H)$ and the control variable is $\sigma \in \Gamma(\mathbb{R}, C(P))$. Following [6], the set of admissible sections is

$$\mathcal{V} = \{(\sigma, \bar{s}) \in \Gamma(C(P) \times_{\mathbb{R}} (P/H)) : \mathrm{im}\ j^1(\sigma, \bar{s}) \subset S\},$$

that is, $\nabla^\sigma \bar{s} = 0$, condition which can be locally written as a system of first order differential equations

$$\left(\frac{d(y^j \circ \bar{s})}{dt} - (A^\alpha \circ \sigma)\left(\Psi^j_\alpha \circ \bar{s}\right)\right) dt \otimes \left(\frac{\partial}{\partial y^j}\right)_{\bar{s}} = 0. \tag{6}$$

Similarly, given an admissible section $(\sigma, \bar{s}) \in \mathcal{V}$, the admissible infinitesimal transformations along $(\sigma, \bar{s})$ are sections

$$(\delta\sigma, \delta\bar{s}) \in \Gamma(\mathbb{R}, \sigma^* V(C(P)) \times \bar{s}^* V(P/H))$$

such that $j^1(\delta\sigma, \delta\bar{s})$ is tangent to the constraint manifold S along $j^1(\sigma, \bar{s})$. We denote by $T_{(\sigma,\bar{s})}\mathcal{V}$ the real vector space of all these infinitesimal transformations. This space is locally characterized by the following set of first order linear differential equations

$$\left[\frac{d(\delta\bar{s})^j}{dt} - (A^\alpha \circ \sigma)\left(\frac{\partial \Psi^j_\alpha}{\partial y^k} \circ \bar{s}\right)(\delta\bar{s})^k - (\delta\sigma)^\alpha(\Psi^j_\alpha \circ \bar{s})\right] dt \otimes \left(\frac{\partial}{\partial y^j}\right)_{\bar{s}} = 0. \tag{7}$$

The main result of this section is the following:

Theorem 1 $(\delta\sigma, \delta\bar{s}) \in T_{(\sigma,\bar{s})}\mathcal{V}$ *if and only if there exist an infinitesimal gauge transformation* $\eta \in \Gamma(\mathbb{R}, \tilde{\mathfrak{g}})$ *and a* 1*-form* $\omega \in \Gamma(\mathbb{R}, T^*\mathbb{R} \otimes \bar{s}^* \tilde{\mathfrak{h}})$ *such that*

$$\delta\sigma = \nabla^{\sigma}\eta + \omega, \qquad \delta\bar{s} = (\eta_{P/H})_{\bar{s}}, \tag{8}$$

where $\tilde{\mathfrak{h}} \to P/H$ *is the adjoint bundle of the principal bundle* $P \to P/H$.

Proof As the operator $\mathrm{P}_{\bar{s}} : \Gamma(\mathbb{R}, \tilde{\mathfrak{g}}) \to \Gamma(\mathbb{R}, \bar{s}^* V(P/H))$ is surjective, there exists $\eta \in \Gamma(\mathbb{R}, \tilde{\mathfrak{g}})$ such that $\mathrm{P}_{\bar{s}}(\eta) = (\eta_{P/H})_{\bar{s}} = \delta\bar{s}$. From (3) and $\eta = \eta^{\alpha}\tilde{B}_{\alpha}, \eta^{\alpha} \in C^{\infty}(\mathbb{R})$, we locally have

$$\delta\bar{s} = (\eta_{P/H})_{\bar{s}} = \eta^{\alpha}(\Psi_{\alpha}^{j} \circ \bar{s}) \left(\frac{\partial}{\partial y^{j}}\right)_{\bar{s}}.$$

If we substitute this expression in (7), we have

$$\begin{aligned}
0 &= \left[\frac{d(\eta^{\alpha}(\Psi_{\alpha}^{j}\circ\bar{s}))}{dt} - (A^{\alpha}\circ\sigma)\left(\frac{\partial\Psi_{\alpha}^{j}}{\partial y^{k}}\circ\bar{s}\right)\eta^{\beta}(\Psi_{\beta}^{k}\circ\bar{s}) - (\delta\sigma)^{\alpha}(\Psi_{\alpha}^{j}\circ\bar{s})\right] dt \otimes \left(\frac{\partial}{\partial y^{j}}\right)_{\bar{s}} \\
&= \left[\frac{d\eta^{\alpha}}{dt}\left(\Psi_{\alpha}^{j}\circ\bar{s}\right) + \eta^{\alpha}\left(\frac{\partial\Psi_{\alpha}^{j}}{\partial y^{k}}\circ\bar{s}\right)\frac{d(y^{k}\circ\bar{s})}{dt} - (A^{\alpha}\circ\sigma)\eta^{\beta}\left(\frac{\partial\Psi_{\alpha}^{j}}{\partial y^{k}}\circ\bar{s}\right)(\Psi_{\beta}^{k}\circ\bar{s})\right. \\
&\quad \left. -(\delta\sigma)^{\alpha}(\Psi_{\alpha}^{j}\circ\bar{s})\right] dt \otimes \left(\frac{\partial}{\partial y^{j}}\right)_{\bar{s}} \\
&= \left[\frac{d\eta^{\alpha}}{dt}\left(\Psi_{\alpha}^{j}\circ\bar{s}\right) + \eta^{\alpha}\left(\frac{\partial\Psi_{\alpha}^{j}}{\partial y^{k}}\circ\bar{s}\right)(A^{\beta}\circ\sigma)(\Psi_{\beta}^{k}\circ\bar{s}) - (A^{\alpha}\circ\sigma)\eta^{\beta}\left(\frac{\partial\Psi_{\alpha}^{j}}{\partial y^{k}}\circ\bar{s}\right)(\Psi_{\beta}^{k}\circ\bar{s})\right. \\
&\quad \left. -(\delta\sigma)^{\alpha}(\Psi_{\alpha}^{j}\circ\bar{s})\right] dt \otimes \left(\frac{\partial}{\partial y^{j}}\right)_{\bar{s}} \\
&= \left[\frac{d\eta^{\alpha}}{dt}\left(\Psi_{\alpha}^{j}\circ\bar{s}\right) + \eta^{\beta}(A^{\gamma}\circ\sigma)\left(\frac{\partial\Psi_{\beta}^{j}}{\partial y^{k}}\Psi_{\gamma}^{k} - \frac{\partial\Psi_{\gamma}^{j}}{\partial y^{k}}\Psi_{\beta}^{k}\right)\circ\bar{s}\right. \\
&\quad \left. -(\delta\sigma)^{\alpha}(\Psi_{\alpha}^{j}\circ\bar{s})\right] dt \otimes \left(\frac{\partial}{\partial y^{j}}\right)_{\bar{s}},
\end{aligned}$$

where we have taken into account the constraint condition $\nabla^{\sigma}\bar{s} = 0$ given in (4). On the other hand, from (2), we have

$$[\tilde{B}_{\beta}, \tilde{B}_{\gamma}]_{P/H} = \left(\frac{\partial\Psi_{\beta}^{j}}{\partial y^{k}}\Psi_{\gamma}^{k} - \frac{\partial\Psi_{\gamma}^{j}}{\partial y^{k}}\Psi_{\beta}^{k}\right)\frac{\partial}{\partial y^{j}},$$

so that

$$\begin{aligned}
0 &= \left(\left(\frac{d\eta^{\alpha}}{dt}\tilde{B}_{\alpha}\right)_{P/H} - \eta^{\beta}(A^{\gamma}\circ\sigma)[\tilde{B}_{\beta}, \tilde{B}_{\gamma}]_{P/H}\right)_{\bar{s}} \otimes dt - \left((\delta\sigma)^{\alpha}\tilde{B}_{\alpha}\right)_{P/H} \otimes dt \\
&= \mathrm{P}_{\bar{s}}(\nabla^{\sigma}\eta - \delta\sigma).
\end{aligned}$$

Hence, $\nabla^{\sigma}\eta - \delta\sigma \in \ker \mathrm{P}_{\bar{s}}$, that is

$$\omega = \nabla^{\sigma}\eta - \delta\sigma \in \Gamma(\mathbb{R}, T^*\mathbb{R}\otimes\bar{s}^*\tilde{\mathfrak{h}}).$$

Conversely, it is clear that any choice of η and ω gives a variation $(\delta\sigma, \delta\bar{s}) \in T_{(\sigma,\bar{s})}\mathscr{V}$.

Corollary 1 *The admissible infinitesimal transformations along an admissible section* $(\sigma, \bar{s}) \in \mathscr{V}$ *of the Euler–Poincaré reduction coincide with the subspace of* $T_{(\sigma,\bar{s})}\mathscr{V}$ *of those* $(\delta\sigma, \delta\bar{s})$ *such that Eq. (8) are satisfied with*

$$\omega = \nabla^{\sigma} \varsigma, \qquad \varsigma \in \Gamma(\mathbb{R}, \bar{s}^{*}\tilde{\mathfrak{h}}). \tag{9}$$

Proof For those transformation there must exist $\eta' \in \Gamma(\mathbb{R}, \tilde{\mathfrak{g}})$ such that

$$\delta\sigma = \nabla^{\sigma}\eta + \omega = \nabla^{\sigma}\eta', \qquad \delta\bar{s} = (\eta_{P/H})_{\bar{s}} = (\eta'_{P/H})_{\bar{s}},$$

and then, $\varsigma = \eta' - \eta \in \Gamma(\mathbb{R}, \bar{s}^{*}\tilde{\mathfrak{h}})$ satisfies $\nabla^{\sigma}\varsigma = \omega$ and conversely.

As a consequence of this Corollary, the critical section of our optimal control problem are also solutions of the Euler–Poincaré equations, but the converse is not true in general. We now explore with more detail this fact on the set of solutions of the optimal control problem obtained through the method of Lagrange multipliers.

4 The Rule of Lagrange Multipliers

Let E_Y be the induced fiber bundle over $Y = C(P) \times_{\mathbb{R}} (P/H)$ from the vector bundle $E = T^{*}\mathbb{R} \otimes_{P/H} V(P/H)$ and E_{J^1Y} the induced bundle over J^1Y from E_Y. The constraint submanifold $S \subset J^1Y$ of the described above can be seen as the preimage of zero of the section $\Phi \in \Gamma(J^1Y, E_{J^1Y})$ defined as

$$\begin{aligned} \Phi \colon J^1Y &\longrightarrow E_{J^1Y} \\ j^1_x(\sigma, \bar{s}) &\mapsto (\nabla^{\alpha}\bar{s})_t. \end{aligned}$$

Following [6] and denoting for simplicity by E all these induced vector bundles, we consider the free variational problem defined on $J^1(Y \times_Y E^{*})$ by the Lagrangian density

$$\hat{l}dt = (l + \lambda \circ \Phi)dt,$$

where $\lambda \in \Gamma(J^1(Y \times_Y E^{*}), E^{*})$ is the section induced by the trivial section

$$\lambda(j^1_t(\sigma, \bar{s}), e^{*}_{(\sigma(t),\bar{s}(t))}) = e^{*}_{(\sigma(t),\bar{s}(t))},$$

and $\circ$ is the duality bilinear product.

Locally, if λ_i, $1 \leq i \leq r$, are the induced coordinates in E^{*} by the coordinates y^i, $1 \leq i \leq r$, of G/H, we have

$$\hat{l} = l(t, A^{\alpha}, y^i) + \sum_{i=1}^{r} \lambda_i \left(y^i - \sum_{\alpha=1}^{n} A^{\alpha} \Psi_{\alpha i}(y^l) \right). \tag{10}$$

As it is well known, the critical sections $(\sigma, \bar{s})$ of this problem are "regular solutions" of the optimal control problem under study (for the Lagrange multipliers rule, see [5, 6]). Under this perspective we have the following result:

Theorem 2 *The Euler–Lagrange equations of the variational problem defined on* $J^1(Y \times_Y E^*)$ *by the Lagrangian density* $\hat{l}dt = (l + \lambda \circ \Phi)dt$ *are*

$$\nabla^\sigma \bar{s} = 0, \tag{11}$$

$$\frac{\delta l}{\delta \sigma} - \mathrm{P}_{\bar{s}}^{+}\lambda = 0, \tag{12}$$

$$\mathrm{P}_{\bar{s}}^{+}\frac{\delta l}{\delta \bar{s}} - \mathrm{div}^\sigma \mathrm{P}_{\bar{s}}^{+}\lambda = 0, \tag{13}$$

where $(\sigma, \bar{s}, \lambda) \in \Gamma(R, Y \times_Y E^*)$, $\mathrm{P}_{\bar{s}}^{+} : \Gamma(\mathbb{R}, \bar{s}^*V^*(P/H)) \to \Gamma(\mathbb{R}, T\mathbb{R} \otimes \tilde{\mathfrak{g}}^*)$ *is the adjoint operator of* $\mathrm{P}_{\bar{s}} : \Gamma(\mathbb{R}, T^*\mathbb{R} \otimes \tilde{\mathfrak{g}}) \to \Gamma(\mathbb{R}, \bar{s}^*V(P/H))$ *and* $\delta l/\delta\sigma \in \Gamma(\mathbb{R}, T\mathbb{R} \otimes \tilde{\mathfrak{g}}^*)$, $\delta l/\delta\bar{s} \in \Gamma(\mathbb{R}, \bar{s}^*V(P/H))$ *are the vertical differential of* l *for* σ *and* $\bar{s}$ *respectively.*

Proof According to (10), the local expressions of the Euler–Lagrange equations are

$$\frac{dy^j}{dt} - \sum_\alpha A^\alpha \Psi_{\alpha j} = 0, \tag{14}$$

$$\frac{\partial l}{\partial A^\alpha} - \sum_i \lambda_i \Psi_{\alpha i} = 0, \tag{15}$$

$$\frac{\partial l}{\partial y^j} - \frac{d\lambda_j}{dt} - \sum_{i,\alpha} A^\alpha \frac{\partial \Psi_{\alpha i}}{\partial y^j}\lambda_i = 0, \tag{16}$$

$1 \le i, j \le r$, $1 \le \alpha \le n$. Equation (14) is the local expression of (11). With respect to Eq. (15), it can be written as

$$\begin{aligned} 0 &= \left\langle \frac{\partial l}{\partial A^\beta}\tilde{B}_\beta^* \otimes \frac{\partial}{\partial t}, \tilde{B}_\alpha \otimes dt \right\rangle - \sum_i \left\langle \lambda_i dy^i \otimes \frac{\partial}{\partial t}, (\tilde{B}_\alpha)_{P/H} y^i \frac{\partial}{\partial y^i} \otimes dt \right\rangle \\ &= \left\langle \frac{\delta l}{\delta \sigma}, \tilde{B}_\alpha \otimes dt \right\rangle - \left\langle \lambda, \mathrm{P}_{\bar{s}}(\tilde{B}_\alpha) \otimes dt \right\rangle = \left\langle \frac{\delta l}{\delta \sigma} - \mathrm{P}_{\bar{s}}^{+}\lambda, \tilde{B}_\alpha \otimes dt \right\rangle, \end{aligned}$$

which gives (12) as $\tilde{B}_\alpha$ is arbitrary. Finally, for (16), taking into account that $\mathrm{P}_{\bar{s}}$: $\Gamma(\mathbb{R}, T^*\mathbb{R} \otimes \tilde{\mathfrak{g}}) \to \Gamma(\mathbb{R}, \bar{s}^*V(P/H))$ is surjective, it is equivalent to

$$\sum_\beta \left[\sum_j \left(\frac{\partial l}{\partial y^j} - \frac{d\lambda_j}{dt} - \sum_\alpha A^\alpha \frac{\partial \Psi_{\alpha i}}{\partial y^j}\lambda_i \right) \Psi_{\beta j} \right] \eta^\beta = 0,$$

for any $\eta = \eta^\alpha \tilde{B}_\alpha$, that is,

$$\sum_j \frac{\partial l}{\partial y^j} \Psi_{\beta j} - \sum_j \left(\frac{d\lambda_j}{dt} \Psi_{\beta j} - \sum_\alpha A^\alpha \frac{\partial \Psi_{\alpha i}}{\partial y^j} \lambda_i \Psi_{\beta j} \right) = 0,$$

for $1 \leq \beta \leq n$. The first term in this expression is

$$\begin{aligned} \sum_j \frac{\partial l}{\partial y^j} \Psi_{\beta j} &= \sum_j \left\langle \frac{\partial l}{\partial y^j} dy^j, (\tilde{B}_\alpha)_{P/H} y^i \frac{\partial}{\partial y^i} \right\rangle \\ &= \left\langle \frac{\delta l}{\delta \bar{s}}, \mathrm{P}_{\bar{s}} \left(\tilde{B}_\beta \right) \right\rangle = \left\langle \mathrm{P}_{\bar{s}}^{+} \frac{\delta l}{\delta \bar{s}}, \tilde{B}_\beta \right\rangle. \end{aligned}$$

With respect to the second term, a computation similar to that of Theorem 1 gives

$$\begin{aligned} &\sum_j \left(\frac{d\lambda_j}{dt} \Psi_{\beta j} - \sum_\alpha A^\alpha \frac{\partial \Psi_{\alpha i}}{\partial y^j} \lambda_i \Psi_{\beta j} \right) \\ &= \sum_j \left[\frac{d}{dt} \left(\lambda_j (\tilde{B}_\beta)_{P/H} y^j \right) - \sum_\alpha \lambda_j A^\alpha [\tilde{B}_\beta, \tilde{B}_\alpha]_{P/H} y^j \right] \\ &= \mathrm{div} \langle \lambda, \mathrm{P}_{\bar{s}} \tilde{B}_\beta \rangle - \left\langle \lambda, \mathrm{P}_{\bar{s}} \left(\nabla^\sigma \tilde{B}_\beta \right) \right\rangle \\ &= \mathrm{div} \langle \mathrm{P}_{\bar{s}}^{+} \lambda, \tilde{B}_\beta \rangle - \langle \mathrm{P}_{\bar{s}}^{+} \lambda, \nabla^\sigma \tilde{B}_\beta \rangle \\ &= \langle \mathrm{div}^\sigma (\mathrm{P}_{\bar{s}}^{+} \lambda), \tilde{B}_\beta \rangle. \end{aligned}$$

As $\tilde{B}_\beta$ is arbitrary, we get (13).

Given a solution $(\sigma, \bar{s}, \lambda) \in \Gamma(\mathbb{R}, Y \times_Y E^*)$ of the Euler–Lagrange equations (11)–(13), we can take the divergence operator in the second and substitute in the third so that we have

$$0 = \mathrm{div}^\sigma \frac{\delta l}{\delta \sigma} - \mathrm{div}^\sigma (\mathrm{P}_{\bar{s}}^{+} \lambda) = \mathrm{div}^\sigma \left(\frac{\delta l}{\delta \sigma} \right) - \mathrm{P}_{\bar{s}}^{+} \frac{\delta 1}{\delta \bar{s}},$$

which, together with $\nabla^\sigma \bar{s} = 0$, are the equations of the Euler–Poincaré reduction (see [3], Sect. 3.2). We thus have that any solution of the optimal control problem obtained through the Lagrange multipliers is also a solution of the Euler–Poincaré reduction. The converse need not be true.

5 Hamilton–Cartan–Pontryagin Formulation

Taking into account (10), the local expression of the Cartan form (see [4]) of the Lagrangian density $\hat{l}dt$ reads

$$\begin{aligned}\Theta_{\hat{l}dt} &= \sum_i \frac{\partial \hat{l}}{\partial \dot{y}^i}\left(dy^i - \dot{y}^i dt\right) + \hat{l}dt \\ &= \sum_i \lambda_i \left(dy^i - \dot{y}^i dt\right) + \left(l + \sum_i \lambda_i \left(\dot{y}^i - \sum_\alpha A^\alpha \Psi_{\alpha i}\right)\right) dt \\ &= \sum_i \lambda_i dy^i - H dt,\end{aligned}$$

where

$$\mathrm{H} = \sum_{i\alpha} \lambda_i A^\alpha \Psi_{\alpha i} - l(t, A^\alpha, y^j) \tag{17}$$

is the Pontryagin Hamiltonian of our problem. From here we see that the Cartan form $\Theta_{\hat{l}dt}$ is projectable to $Y \times_Y E^*$.

On the other hand, Euler–Lagrange equation (12) defines a closed subset $W \subset Y \times_Y E^*$ fibering over $\mathbb{R}$ for which the following notion of tangent space can be given at $w \in W$

$$T_w W = \{\hat{D}_w \in T_w(Y \times_Y E^*) : \hat{D}_w I_w = 0\},$$

where I_w is the ideal of germs in w of functions in $C^\infty(Y \times_Y E^*)$ vanishing on W.

In terms of the Pontryagin Hamiltonian (17), the local Euler–Lagrange equations (14)–(16) read as

$$\frac{dy^i}{dt} = \frac{\partial \mathrm{H}}{\partial \lambda_i}, \tag{18}$$

$$\frac{\partial \mathrm{H}}{\partial A^\alpha} = 0, \tag{19}$$

$$\frac{d\lambda_i}{dt} = -\frac{\partial \mathrm{H}}{\partial y^i}, \tag{20}$$

with $1 \leq \alpha \leq n$, $1 \leq i \leq r$. In particular, the subset W is the locus where $\partial \mathrm{H}/\partial A^\alpha = 0$, $1 \leq \alpha \leq n$, so that for every $w \in W$ we have

$$T_w W = \{\hat{D}_w \in T_w(Y \times_Y E^*) : \hat{D}_w \left(\partial H/\partial A^\alpha\right) = 0, 1 \leq \alpha \leq n\}.$$

A key point in our results relies in the following condition of regularity.

Definition 1 A Lagrangian density $\hat{l}dt$ is said to be regular if for every $w \in W \subset Y \times_Y E^*$, the polarity $\hat{D}_w \mapsto i_{\hat{D}_w} d\Theta_{\hat{l}dt}$ is injective on the set of vertical vectors $\hat{D}_w \in T_w W$.

If

$$\hat{D}_w = a^i \left(\frac{\partial}{\partial y^i}\right)_w + b^\alpha \left(\frac{\partial}{\partial A^\alpha}\right)_w + c_i \left(\frac{\partial}{\partial \lambda_i}\right)_w$$

belongs to the kernel of the polarity, that is

$$\begin{aligned}0 = i_{\hat{D}_w} d\Theta_{\hat{l}dty} &= i_{\hat{D}_w}\left(\sum_i \lambda_i \wedge dy^i - d\mathrm{H} \wedge dt\right) \\ &= \sum_i \left(c_i dy^i - a^i d\lambda_i\right) - (\hat{D}_w \mathrm{H})dt,\end{aligned}$$

and then $a^i = c_i = 0$ and $\hat{D}_w(\mathrm{H}) = \sum_\alpha b^\alpha \left(\partial H/\partial A^\alpha\right)(w) = 0$. Substituting in $\hat{D}_w\left(\partial \mathrm{H}/\partial A^\alpha\right) = 0$ we have

$$\sum_\beta \left(\frac{\partial^2 \mathrm{H}}{\partial A^\alpha \partial A^\beta}\right)(w) b^\beta = 0, \qquad 1 \leq \alpha \leq n,$$

and the polarity is injective if and only if

$$\det\left(\frac{\partial^2 \mathrm{H}}{\partial A^\alpha \partial A^\beta}\right) = \det\left(-\frac{\partial^2 l}{\partial A^\alpha \partial A^\beta}\right) \neq 0 \tag{21}$$

along $W \subset Y \times_Y E^*$. In particular, condition (21) implies that W is a submanifold of dimension $2r + 1$ with local coordinates (t, y^i, λ_i), $1 \leq i \leq r$, and parametric equations

$$t = t, \quad y^i = y^i, \quad \lambda_i = \lambda_i, \quad A^\alpha = A^\alpha(t, y^i, \lambda_i),$$

by virtue of the Implicit function Theorem applied to the constraints $\partial \mathrm{H}/\partial A^\alpha = 0$, $1 \leq \alpha \leq n$.

From this fact we have that the next two 1-forms

$$\left(dt, \Theta_{\hat{l}dt}|_W = \sum_{i=1}^r \lambda_i dy^i - \mathrm{H}|_W dt\right)$$

on $W \subset Y \times_Y E^*$ define a cosymplectic structure with which the Euler–Lagrange equations of the Lagrangian density $\hat{l}dt$ in $J^1(Y \times_Y E^*)$ can be characterized as follows:

Theorem 3 (Hamilton–Cartan–Pontryagin formulation) *There exists a unique vector field $\hat{D}$ in the bundle $W \to \mathbb{R}$ such that*

$$i_{\hat{D}} d\Theta_{\hat{l}dt}|_W = 0, \qquad \hat{D}(t) = 1.$$

The integral curves of $\hat{D}$ are the solutions of the Euler–Lagrange equations (11)–(13).

6 Application to the Dynamic of the Heavy Top

In this case $P = \mathbb{R} \times SO(3) \to \mathbb{R}$ and $H = SO(2) \subset SO(3)$ and $SO(3)/SO(2) = S^2$. The reduced fiber bundle $Y = C(P) \times_{\mathbb{R}} (P/H)$ can be written as the fiber product

$$(\mathbb{R} \times \mathbb{T}^*\mathbb{R} \otimes \mathfrak{so}(3)) \times_{\mathbb{R}} (\mathbb{R} \times S^2) = \mathbb{R} \times (\mathfrak{so}(3) \times S^2) \to \mathbb{R}.$$

Therefore, sections $(\sigma, \bar{s})$ of this bundle are of the type

$$(\sigma, \bar{s})(t) = (t, \Omega(t), \Lambda(t))$$

for curves

$$\Omega : \mathbb{R} \to \mathfrak{so}(3) = \mathbb{R}^3,$$
$$\Lambda : \mathbb{R} \to \mathbb{S}^2 \subset \mathbb{R}^3,$$

where the identification $\mathfrak{so}(3) = \mathbb{R}^3$ is the standard one

$$(a, b, c) \mapsto \begin{pmatrix} 0 & c & -b \\ -c & 0 & a \\ b & -a & 0 \end{pmatrix}.$$

Following [3], Sect. 5, the constraint $\nabla^\sigma \bar{s} = 0$ reads as

$$\frac{d\Lambda}{dt}(t) + \Omega(t) \times \Lambda(t) = 0, \tag{22}$$

where $\times$ stands for the cross product in $\mathbb{R}^3$.

On the other hand, taking into account that $\Gamma(\mathbb{R}, \widetilde{\mathfrak{so}}(3)) = C^\infty(\mathbb{R}, \mathfrak{so}(3))$ and $\Gamma(\mathbb{R}, \bar{s}^* V(P/H)) = \Gamma(\mathbb{R}, \Lambda^* T S^2)$, the operators $\mathrm{P}_\Lambda : C^\infty(\mathbb{R}, \mathfrak{so}(3) = \mathbb{R}^3) \to \Gamma(\mathbb{R}, \Lambda T S^2)$ and $\mathrm{P}_\Lambda^+ : \Gamma(\mathbb{R}, \Lambda^* T^* S^2) \to C^\infty(\mathbb{R}, \mathfrak{so}(3)^* = \mathbb{R}^3)$ are

$$\mathrm{P}_\Lambda(\eta) = \Lambda \times \eta, \qquad \mathrm{P}_\Lambda^+ \Upsilon = -\Lambda \times \Upsilon. \tag{23}$$

In particular

$$\ker \mathrm{P}_\Lambda = \{a(t)\Lambda(t) : a(t) \in C^\infty(\mathbb{R})\}.$$

Therefore, the 1-forms ω mentioned in Theorem 1 have the expression

$$\omega = a(t)\Lambda(t) \otimes dt, \qquad a(t) \in C^\infty(\mathbb{R}). \tag{24}$$

Taking the infinitesimal variations (8) for the Lagrangian density $l dt$ as $\eta \in \Gamma(\mathbb{R}, \widetilde{\mathfrak{so}}(3))$, with compact support, and $\omega = 0$, we obtain the Euler–Poincaré equations. For $\eta = 0$ and ω as in (24), with a compactly supported, we have the additional equation

$$\left\langle \frac{\delta l}{\delta \sigma}, \Lambda \right\rangle = 0, \tag{25}$$

which gives the zero level set of the vertical component of the angular momentum of the heavy top. Recall that this component is a first integral.

With respect to Corollary 1, Eq. (9) has always a solution. In fact, given $\omega = a(t)\Lambda(t) \otimes dt$, if $\zeta = b(t)\Lambda(t)$ satisfies $\nabla^\sigma \zeta = \omega$ we have

$$\left(\frac{d(b\Lambda)}{dt} + \Omega \times b\Lambda \right) \otimes dt = (a\Lambda) \otimes dt,$$

and taking into account (22)

$$\frac{db}{dt}(t) = a(t), \quad \text{that is, } b(t) = \int a(t)dt + K,$$

for certain constant $K \in \mathbb{R}$.

Hence, the admissible infinitesimal variations of the Euler–Poincaré reduction of the heavy top coincide with those of the associated optimal control problem. But this does not mean that both problems define the same critical curves because solutions of the optimal control problem satisfy the additional Eq. (25). The point is that the coincidence in the set of admissible infinitesimal variations becomes a proper inclusion when taking compactly supported variations to define the critical curves.

From (22) and (23), Eqs. (11)–(13) obtained as the result of the rule of Lagrange multipliers for the heavy top read

$$\frac{d\Lambda}{dt} + \Omega \times \Lambda = 0, \tag{26}$$

$$\frac{\delta l}{\delta \Omega} + (\Lambda \times \lambda) \otimes dt = 0, \tag{27}$$

$$-\Lambda \times \frac{\delta l}{\delta \Lambda} + \frac{d}{dt}(\Lambda \times \lambda) + \Omega \times (\Lambda \times \lambda) = 0, \tag{28}$$

where we have used the expression for the divergence operator div^σ

$$\mathrm{div}^\sigma \left(\xi \otimes \frac{\partial}{\partial t} \right) = \frac{d\xi}{dt} + \Omega \times \xi, \qquad \xi \in C^\infty(\mathbb{R}, \mathfrak{so}(3)^* = \mathbb{R}^3).$$

Substituting Eq. (27) in Eq. (28), we get

$$\frac{d}{dt}\left(\frac{\delta l}{\delta \Omega} \right) + \Omega \times \frac{\delta l}{\delta \Omega} + \Lambda \times \frac{\delta l}{\delta \Lambda} = 0$$

which, together with (26), are the Euler–Poincaré equations of the heavy top.

On the other hand, multiplying Eq. (27) by Λ, we obtain Eq. (25) and conversely. Therefore, the solutions of the optimal control problem defined by the rule of Lagrange multipliers are the solutions of the Euler–Poincaré equations lying on the zero level set of the first integral $\langle \delta l/\delta \Omega, \Lambda \rangle$, the vertical component of the angular momentum.

Finally, due to the identification $Y = \mathbb{R} \times (\mathfrak{so}(3) \times S^2) \to \mathbb{R}$ we have

$$Y \times_Y E^* = \mathbb{R} \times (\mathfrak{so}(3) \times T^*S^2) \to \mathbb{R},$$

so that, according to the local expression of Sect. 5, the Cartan 1-form $\Theta_{\hat{l}dt}$ projects in this case to the following 1-form on $\mathbb{R} \times (\mathfrak{so}(3) \times T^*S^2)$

$$\Theta_{\hat{l}dt} = \theta - \mathrm{H}dt,$$

where θ is the Liouville 1-form on T^*S^2 and H is the function

$$\mathrm{H}(t, \Omega, \lambda) = \langle \lambda, \mathrm{P}_\Lambda(\Omega) \rangle - l = \langle \lambda, \Lambda \times \Omega \rangle - l.$$

Furthermore, if the problem is regular in the sense of Definition 1, we can locally solve for Ω in the equation $\partial H/\partial \Omega = 0$ defining the submanifold

$$W \subset \mathbb{R} \times (\mathfrak{so}(3) \times T^*S^2),$$

as $\Omega = \Omega(t, \Lambda, \lambda)$, and we have a local diffeomorphism $\Psi : \mathbb{R} \times T^*S^2 \to W$ such that

$$\Psi^* \Theta_{\hat{l}dt}|_W = \theta - \mathrm{H}(t, \Omega(t, \Lambda, \lambda), \lambda)dt.$$

The case where the diffeomorphism is global (that is, hyper-regularity) is of special interest. This situation occurs for the standard Lagrangian of the heavy top

$$l = \tfrac{1}{2}\langle \mathbb{I}\Omega, \Omega \rangle - mg\langle \Lambda, \chi \rangle,$$

where $\mathbb{I}$ is the inertia tensor of the body, $\langle \cdot, \cdot \rangle$ is the standard inner product in $\mathbb{R}^3$, and χ is the vector joining the fixed point of the heavy top with its center of mass. In this case the equation $\partial \mathrm{H}/\partial \Omega = 0$ reads $\mathbb{I}\Omega = \lambda \times \Lambda$, and hence $\Omega = \mathbb{I}^{-1}(\lambda \times \Lambda)$. We can thus state Theorem 3 as follows:

Solutions of the Euler–Lagrange equations (26)–(28) of the heavy top are the integral curves of the Hamiltonian vector field D_{H} on $(T^*S^2, d\theta)$ defined by the Hamiltonian

$$H = \tfrac{1}{2}\langle \mathbb{I}^{-1}(\lambda \times \Lambda), (\lambda \times \Lambda \rangle - mg\langle \Lambda, \chi \rangle,$$

where $d\theta$ is the canonical symplectic form of the cotangent bundle.

Acknowledgments Both authors have been partially supported by project MTM-2010-19111 (Ministerio de Ciencia e Innovación, Spain).

References

1. Bloch, A.M.: Nonholonomic Mechanics and Control, Interdisciplinary Applied Mathematics, Systems and Control, vol. 24. Springer, Berlin (1988)
2. Castrillón López, M., García Pérez, P.L., Rodrigo, C.: Euler-Poincaré reduction in principal fiber bundles and the problem of Lagrange. Differ. Geom. Appl. **25**(6), 585–593 (2007)
3. Castrillón López, M., García Pérez, P.L., Rodrigo, C.: Euler-Poincaré reduction in principal bundles by a subgroup of the structure group. J. Geom. Phys. **74**, 352–369 (2013)
4. García, P.L.: The Poincaré-Cartan Invariant in the Calculus of Variations. In: Symposia Mathematica, vol. 14, pp. 219–249. Academic Press, London (1974)
5. García, P.L.: Sobre la regularidad en los problemas de Lagrange y de control óptimo, Actas del Simposio en memoria de J.B. Sancho Guimerá. Ediciones de la Universidad de Salamanca (in press)
6. García, P.L., García, A., Rodrigo, C.: Cartan forms for first order constrained variational problems. J. Geom. Phys. **56**, 571–610 (2006)
7. Koszul, J.L.: Lectures on fibers bundles and differential geometry. In: Lectures in Mathematics, vol. 20. Tata Institute of Fundamental Research, Bombay (1965)

Morse Families and Lagrangian Submanifolds

Marco Castrillón López and Tudor S. Ratiu

Dedicated to Jaime Muñoz Masqué on the occasion of his 65th birthday.

Abstract This short note presents a comprehensive and pedagogical study of the results in [6] on Lagrangian submanifolds in cotangent bundles T^*X defined by Morse families $S : B \to \mathbb{R}$ for arbitrary submersions $B \to X$.

Keywords Symplectic manifold · Lagrangian submanifold · Morse family

1 Introduction

Lagrangian submanifolds play an essential role in the study of symplectic manifolds, either from a pure mathematical point of view or, in geometric mechanics, when applied to the Hamiltonian formulation of the equations of motion. With respect to the latter, Lagrangian submanifolds naturally appear, for example, as singularities

M. Castrillón López (✉)
ICMAT(CSIC-UAM-UC3M-UCM), Departamento de Geometría y Topología,
Facultad de CC. Matemáticas, Universidad Complutense de Madrid,
28040 Madrid, Spain
e-mail: mcastri@mat.ucm.es

T.S. Ratiu
Department of Mathematics, Shanghai Jiao Tong University,
800 Dongchuan Road, Minhang District, 200240 Shanghai, China
e-mail: ratiu@sjtu.edu.cn

T.S. Ratiu
Séction de Mathématiques, École Polythechnique Fédérale de Lausanne,
1015 Lausanne, Switzerland
e-mail: tudor.ratiu@epfl.ch

M. Castrillón López et al. (eds.), *Geometry, Algebra and Applications: From Mechanics to Cryptography*, Springer Proceedings in Mathematics & Statistics 161, DOI 10.1007/978-3-319-32085-4_6

in ray optics or the level sets of the functions in involution for integrable systems in the Liouville sense. With respect to the former, there are many concepts that can be regarded as Lagrangian submanifolds of cotangent bundles endowed with the canonical symplectic form, such as symplectomorphisms or closed 1-forms, identified with their graphs. From this point of view, the specific case of exact 1-forms is particularly interesting and is exploited in the geometric formulation of the Hamilton-Jacobi theory.

Recent renewed interest in the Cotangent Bundle Reduction Theorem and its applications to mechanical systems (see, e.g., [1, Theorem 4.3.3 and Sect. 4.5]) and the structure of coadjoint orbits (see, e.g., [4]), a reduction theorem for Hamilton-Jacobi theory by a group of point transformations under certain invariance hypotheses (see [4]), and our own attempts to find a general reduction theory for the Hamilton-Jacobi equations, led us to review some very interesting old ideas of Alan Weinstein presented in his well-known lectures [6]. The most general Cotangent Bundle Reduction Theorem can be found in Lecture 6 of these notes; it is a vast generalization of the usual Cotangent Bundle Reduction Theorem at the zero value of the momentum map. He also describes there families $S : N \times X \to \mathbb{R}$ of functions, where N is a manifold of labels, when studying Lagrangian submanifolds of T^*X. In fact, the setup is completely general and is given by a function S is defined on a manifold B and an arbitrary submersion $B \to X$. Under suitable topological conditions, he calls these functions S Morse families and studies their properties. Undoubtedly, the results of [6] have been the inspiring source for many subsequent works and are a classical contribution to symplectic geometry.

The goal of this short note is to give a comprehensive and pedagogical presentation of the results in [6, Lecture 6] concerning reduction by a coisotropic regular foliation and Morse families. We believe that a more elaborate, complete, and self-contained presentation of these, apparently forgotten, ideas and proofs in [6], is helpful and may turn out to be crucial for future investigations in Lagrangian submanifolds and, in particular, the Hamilton-Jacobi theory in the context of reduction by a group of symmetries.

Notations and conventions. Unless otherwise indicated, all objects are smooth. The Einstein summation convention on repeated sub- and super-indices is used. If $E \to Q$ is a vector bundle over the smooth manifold Q and $E^* \to Q$ its dual, $\langle \cdot, \cdot \rangle : E^* \times E \to \mathbb{R}$ denotes the standard fiberwise duality pairing. Given a smooth manifold Q, $\tau_Q : TQ \to Q$ and $\pi_Q : T^*Q \to Q$ denote its tangent and cotangent bundles. If $(q^1, \ldots, q^n)$ are local coordinates on Q, the naturally induced coordinates on TQ and T^*Q are denoted by $(q^1, \ldots, q^n, \dot{q}^1, \ldots, \dot{q}^n)$ and $(q^1, \ldots, q^n, p_1, \ldots, p_n)$, respectively, i.e., any tangent vector $v_q \in T_qQ$ is written locally as $v_q = \dot{q}^i \frac{\partial}{\partial q^i}$ and any covector $\alpha_q \in T_q^*Q$ is written locally as $\alpha_q = p_i dq^i$. If Q and P are manifolds and $f : Q \to P$ a smooth map, $Tf : TQ \to TP$ denotes its tangent map, or derivative.

The canonical, or Liouville, one-form θ_Q on T^*Q is defined by $\theta_Q(\alpha_q)\left(V_{\alpha_q}\right) := \left\langle \alpha_q, T_{\alpha_q}\pi_Q\left(V_{\alpha_q}\right)\right\rangle$ for any $q \in Q$, $\alpha_q \in T_q^*Q$, and $V_{\alpha_q} \in T_{\alpha_q}(T^*Q)$. The canonical symplectic two-form on T^*Q is defined by $\omega_Q := -\mathbf{d}\theta_Q$, were $\mathbf{d}$ denotes the exterior derivative. In standard cotangent bundle coordinates, we have $\theta_Q = p_i dq^i$ and $\omega_Q = dq^i \wedge dp_i$, where $\wedge$ is the exterior product on forms (with Bourbaki conventions).

2 Coisotropic Reduction of Conormal Bundles

We recall standard terminology from symplectic geometry. If (P, ω) is a symplectic manifold (i.e., the two-form ω on P is closed and non-degenerate) and $E \subset TP$ is a vector subbundle, its *ω-orthogonal vector subbundle* is defined by $E^{\perp} := \{v \in T_pP \mid \omega(p)(u, v) = 0,\ \forall u \in E_p,\ \forall p \in P\}$. The vector subbundle $E \subset TP$ is called *isotropic* (*coisotropic*), if $E \subseteq E^{\perp}$ ($E \supseteq E^{\perp}$). The vector subbundle E is called *Lagrangian*, if it is isotropic and has an isotropic complementary vector subbundle F, i.e., $F \subseteq F^{\perp}$ and $TP = E \oplus F$. Thus, E is Lagrangian if and only if $E = E^{\perp}$ if and only if E is isotropic and its rank is half the dimension of P. In addition, its isotropic complement F is actually Lagrangian. The vector subbundle E is *symplectic*, if ω restricted to $E \times E$ is nondegenerate. Thus, E is symplectic if and only if $E \oplus E^{\perp} = TP$. The same terminology is used for vector subbundles of TP restricted to a submanifold of P.

A submanifold $M \subset P$ is called *isotropic* (*coisotropic, Lagrangian, symplectic*) if its tangent bundle is isotropic (coisotropic, Lagrangian, symplectic) in the restriction $(TP)_M$ of the tangent bundle TP to M. For example, a submanifold M is isotropic if and only if $\iota^*\omega = 0$, where $\iota : M \hookrightarrow P$ is the inclusion. A submanifold M of P is Lagrangian if and only if $\iota^*\omega = 0$ and $\dim M = \frac{1}{2}\dim P$.

Let X be a manifold and $\pi : B \to X$ a smooth submersion (possibly a fiber bundle). Since π is a submersion, it is open and hence $\pi(B)$ is an open subset of X. Let $\pi_B : T^*B \to B$ and $\pi_X : T^*X \to X$ be the cotangent bundle projections. The *conormal bundle to the fibers* $\pi_B|_{N_\pi} : N_\pi := (\ker T\pi)^\circ \to B$ is the vector subbundle of T^*B consisting of all covectors annihilating $\ker T\pi$, i.e., the fiber $(N_\pi)_b$ of N_π at $b \in B$ is the vector subspace $(N_\pi)_b := \{\alpha_b \in T^*_bB \mid \langle \alpha_b, v_b \rangle = 0, \forall v_b \in \ker T_b\pi\}$. The upper circle on a vector subspace denotes its annihilator in the dual of the ambient vector space.

Lemma 1 *The conormal bundle to the fibers $N_\pi \subset T^*B$ is a coisotropic submanifold with respect to the canonical symplectic form on T^*B.*

Proof Let $n = \dim X$, $n + k = \dim B$. Since π is a surjective submersion, it is locally expressed as a projection, i.e., around every point $b \in B$ there are coordinates $(x^1, \ldots, x^n, a^1, \ldots, a^k)$ on B such that π has the expression

$$\pi(x^1, \ldots, x^n, a^1, \ldots, a^k) = (x^1, \ldots, x^n).$$

In these coordinates, we express an arbitrary covector $\alpha_b \in T^*_bB$ as

$$\alpha_b = p_1 dx^1 + \cdots + p_n dx^n + \alpha_1 da^1 + \cdots + \alpha_k da^k, \quad p_1, \ldots, p_n, \alpha_1, \ldots, \alpha_k \in \mathbb{R}.$$

In these coordinates,

$$\ker T\pi = \text{span}\left\{\frac{\partial}{\partial a^1}, \ldots, \frac{\partial}{\partial a^k}\right\},$$

and hence $\alpha_b \in N_\pi$ if and only if

$$\alpha_b = p_1 dx^1 + \cdots + p_n dx^n,$$

that is, N_π is locally defined by the equations $\alpha_1 = \cdots = \alpha_k = 0$, i.e., the local expression of N_π is

$$N_\pi = \left\{(x^1, \ldots, x^n, a^1, \ldots, a^k, p_1, \ldots, p_n, 0, \ldots, 0) \in \mathbb{R}^{n+k}\right\}. \tag{1}$$

This shows that N_π is a submanifold of T^*B of dimension $2n + k$, whose tangent space at any $\alpha_b \in N_\pi$ is expressed locally as

$$T_{\alpha_b} N_\pi = \text{span}\left\{\frac{\partial}{\partial x^1}, \ldots, \frac{\partial}{\partial x^n}, \frac{\partial}{\partial a^1}, \ldots, \frac{\partial}{\partial a^k}, \frac{\partial}{\partial p_1}, \ldots, \frac{\partial}{\partial p_n}\right\}. \tag{2}$$

Let $\omega_B \in \Omega^2(T^*B)$ be the canonical cotangent bundle symplectic form which, in these local coordinates, has the expression

$$\omega_B = dx^1 \wedge dp_1 + \cdots + dx^n \wedge dp_n + da^1 \wedge d\alpha_1 \wedge + \cdots + da^k \wedge d\alpha_k.$$

Using (2), the ω_B-orthogonal complement of $T_{\alpha_b} N_\pi$ (taken fiber-wise) is easily calculated in these local coordinates to be

$$(T_{\alpha_b} N_\pi)^\perp = \text{span}\left\{\frac{\partial}{\partial a^1}, \ldots, \frac{\partial}{\partial a^k}\right\}. \tag{3}$$

Thus, $(T_{\alpha_b} N_\pi)^\perp \subset T_{\alpha_b} N_\pi$, which shows that N_π is coisotropic in T^*B. □

The local expression of $(TN_\pi)^\perp \subset T(T^*B)$ implies that the vector subbundle $(TN_\pi)^\perp$ is integrable. This is a general fact, namely, the tangent bundle of a coisotropic submanifold is an integrable subbundle of the tangent bundle of the ambient symplectic manifold (see, e.g., [6, Lecture 3] or [1, Proposition 5.3.22]). Denote by $\mathcal{N}_\pi^\perp$ the foliation in T^*B defined by the integrable vector subbundle $(TN_\pi)^\perp$. From now on we assume that $\mathcal{N}_\pi^\perp$ is a *regular foliation*, i.e., its space of leaves $N_\pi / \mathcal{N}_\pi^\perp$ is a smooth manifold and the canonical projection $\rho : N_\pi \to N_\pi / \mathcal{N}_\pi^\perp$ is a smooth submersion; this uniquely determines the manifold structure on the space of leaves, assuming it exists.

By (3), in local coordinates, the leaf of the foliation $\mathcal{N}_\pi^\perp$ containing the point

$$(x_0^1, \ldots, x_0^n, a_0^1, \ldots, a_0^k, (p_1)_0, \ldots, (p_n)_0, 0, \ldots, 0) \in N_\pi$$

is given by

$$\{(x_0^1, \ldots, x_0^n, a^1, \ldots, a^k, (p_1)_0, \ldots, (p_n)_0, 0, \ldots, 0) \mid a^1, \ldots, a^k \in \mathbb{R}\}. \tag{4}$$

Therefore, the projection $\rho : N_\pi \to N_\pi/\mathcal{N}_\pi^\perp$ has the local expression

$$\rho(x^1, \ldots, x^n, a^1, \ldots, a^k, p_1, \ldots, p_n, 0, \ldots, 0) = (x^1, \ldots, x^n, p_1, \ldots, p_n) \tag{5}$$

and hence $(x^1, \ldots, x^n, p_1, \ldots, p_n)$ are local coordinates on $N_\pi/\mathcal{N}_\pi^\perp$ (remember that $N_\pi/\mathcal{N}_\pi^\perp$ is, by assumption, a manifold).

By the Coisotropic Reduction Theorem (see, e.g., [6, Lecture 3] or [1, Theorem 5.3.33]), the quotient $N_\pi/\mathcal{N}_\pi^\perp$ has a canonical symplectic form ω_π, uniquely characterized by $\rho^*\omega_\pi = \iota^*\omega_B$, where $\iota : N_\pi \hookrightarrow T^*B$ is the inclusion.

Proposition 1 ([6], Lecture 6) *The following statements hold.*

(i) *Let*

$$\pi^*T^*X = \left\{\left(b, \beta_{\pi(b)}\right) \,\middle|\, b \in B, \beta_{\pi(b)} \in T^*_{\pi(b)}X\right\} \ni (b, \beta_{\pi(b)}) \longmapsto b \in B$$

be the pull-back bundle to B of the cotangent bundle $T^*X \to X$ by π. The map $\Xi : N_\pi \to \pi^*T^*X$ given by $\Xi(\alpha_b) := \left(b, \beta_{\pi(b)}\right)$, where $\alpha_b \in (N_\pi)_b = N_\pi \cap T_b^*B$ and $\beta_{\pi(b)} \in T^*_{\pi(b)}X$ is defined by $\left\langle \beta_{\pi(b)}, T_b\pi(v_b)\right\rangle := \langle \alpha_b, v_b\rangle$ for all $v_b \in T_bB$, is a vector bundle isomorphism. Its inverse $\Xi^{-1} : \pi^*T^*X \to N_\pi$ is $\Xi^{-1}(b, \beta_{\pi(b)}) = \beta_{\pi(b)} \circ T_b\pi$.

(ii) *Define the submersion $\widetilde{\pi} : \pi^*T^*X \to T^*X$ by $\widetilde{\pi}(b, \beta_{\pi(b)}) := \beta_{\pi(b)}$. Then, for any $\alpha_b \in N_\pi$, we have $T_{\alpha_b}\Xi\left(\left(T_{\alpha_b}N_\pi\right)^\perp\right) = \ker T_{\alpha_b}\widetilde{\pi}$; recall that $\left(T_{\alpha_b}N_\pi\right)^\perp$ is the tangent space at α_b to the fiber of the foliation $\mathcal{N}_\pi^\perp$ containing α_b.*

(iii) *The integral leaves of the foliation $\mathcal{N}_\pi^\perp$ in N_π are images under Ξ^{-1} of the connected components of the fibers of $\widetilde{\pi} : \pi^*T^*X \to T^*X$.*

(iv) *The map $\Phi : N_\pi/\mathcal{N}_\pi^\perp \to T^*X$ defined by $\Phi([\alpha_b]) := \widetilde{\pi}(\Xi(\alpha_b))$ is well-defined, a local diffeomorphism, and a symplectic map.*

(v) *The map Φ is surjective if and only if π is surjective. The map Φ is injective if and only if the fibers of π over $\pi(B)$ are connected.*

(vi) *If π is surjective and has connected fibers, then $\Phi : \left(N_\pi/\mathcal{N}_\pi^\perp, \omega_\pi\right) \to (T^*X, \omega_B)$ is a symplectic diffeomorphism.*

The spaces and maps involved in this proposition are summarized in the commutative diagram below. The first vertical arrow is only a surjective submersion, whereas the second, third, and fourth are vector bundle projections.

$$
\begin{array}{ccccccc}
 & & & \overset{\Phi}{\curvearrowright} & & & \\
N/\mathcal{N}_\pi^\perp & \xleftarrow{\ \rho\ } & N_\pi = (\ker T\pi)^\circ & \xrightarrow{\ \Xi\ } & \pi^*T^*X & \xrightarrow{\ \tilde{\pi}\ } & T^*X \\
\downarrow & & \downarrow {\scriptstyle \pi_B|_{N_\pi}} & & \downarrow & & \downarrow {\scriptstyle \pi_X} \\
X \supseteq \pi(B) & \xleftarrow[\ \pi\ ]{} & B & = & B & \xrightarrow[\ \pi\ ]{} & X
\end{array}
\tag{6}
$$

Point (vi) of this proposition is a vast generalization of the Cotangent Bundle Reduction Theorem ([1, Theorem 4.3.3], [4, Chap. 2]) at the zero value of the momentum map. Finding a similar generalization of this theorem for any value of the momentum map would be very interesting and relate to the general reduction procedure for Hamilton–Jacobi theory.

Proof (i) The map $\Xi : N_\pi \to \pi^*T^*X$ is well defined. Indeed, any tangent vector in $T_{\pi(b)}X$ is necessarily of the form $T_b\pi(v_b)$ for some $v_b \in T_bB$ because π is a submersion. If $T_b\pi(v_b) = T_b\pi(v'_b)$ for $v_b, v'_b \in T_bB$, i.e., $v_b - v'_b \in \ker T_b\pi$, then $\langle \alpha_b, v_b - v'_b \rangle = 0$ for any $\alpha_b \in (\ker T_b\pi)^\circ \subset N_\pi$. This shows that $\langle \alpha_b, v_b \rangle = \langle \alpha_b, v'_b \rangle$ thus proving that Ξ is well defined.

We compute the local expression of Ξ. If

$$\alpha_b = (x^1, \ldots, x^n, a^1, \ldots, a^n, p_1, \ldots, p_n, 0, \ldots, 0) = p_1 dx^1 + \cdots + p_n dx^n \in N_\pi,$$

and $\Xi(\alpha_b) = (b, \beta_{\pi(b)})$, with $\beta_{\pi(b)} = (x^1, \ldots, x^n, r_1, \ldots, r_n) = r_1 dx^1 + \cdots + r_n dx^n$, then choosing and arbitrary vector

$$v_b = u^1 \frac{\partial}{\partial x^1} + \cdots + u^n \frac{\partial}{\partial x^n} + v^1 \frac{\partial}{\partial a^1} + \cdots + v^k \frac{\partial}{\partial a^k} \in T_bB,$$

we have

$$r_1 u^1 + \cdots + r_n u^n = \langle \beta_{\pi(b)}, T_b\pi(v_b) \rangle = \langle \alpha_b, v_b \rangle = p_1 u^1 + \cdots + p_n u^n$$

for any $u^1, \ldots, u^n \in \mathbb{R}$, i.e., $\beta_{\pi(b)} = p_1 dx^1 + \cdots + p_n dx^n$. This shows that, choosing the standard cotangent bundle coordinates on T^*X,

$$\Xi(x^1, \ldots, x^n, a^1, \ldots, a^n, p_1, \ldots, p_n) = (x^1, \ldots, x^n, a^1, \ldots, a^n, p_1, \ldots, p_n), \tag{7}$$

that is, Ξ is the identity map in these coordinate systems. Thus, Ξ is smooth and, from the very definition of Ξ, it is a vector bundle morphism.

The smooth vector bundle morphism $\pi^*T^*X \ni (b, \beta_{\pi(b)}) \longmapsto \beta_{\pi(b)} \circ T_b\pi \in N_\pi$ is easily verified to equal $\Xi^{-1} : \pi^*T^*X \to N_\pi$, i.e., $\Xi^{-1}(b, \beta_{\pi(b)}) = \beta_{\pi(b)} \circ T_b\pi$, as claimed in the statement.

(ii) Work with the same coordinate systems $(x^1, \ldots, x^n)$ on X, $(x^1, \ldots, x^n, a^1, \ldots, a^k)$ on B, and $(x^1, \ldots, x^n, a^1, \ldots, a^n, p_1, \ldots, p_n)$ on both $N_\pi \subset T^*B$ and π^*T^*X, as in the proof of (i). In these coordinate systems, the projection $\widetilde{\pi}$ reads

$$(x^1, \ldots, x^n, a^1, \ldots, a^k, p_1, \ldots, p_n) \mapsto (x^1, \ldots, x^n, p_1, \ldots, p_n).$$

On the other hand, the local expression of Ξ is the identity, as was shown in the proof of (i). Thus, taking into account formula (3) for the tangent space of a leaf of $\mathscr{N}_\pi^\perp$ at a point $\alpha_b \in N_\pi$, we conclude that its image under $T_{\alpha_b}\Xi$ is $\operatorname{span}\{\partial/\partial a^1, \ldots, \partial/\partial a^k\}$, which is exactly the kernel of $T_{\alpha_b}\widetilde{\pi}$.

(iii) Recall that Ξ is a vector bundle isomorphism. By (ii), its tangent map is an isomorphism of the vector subbundle $TN_\pi^\perp \subset TN_\pi$, defining the foliation $\mathscr{N}_\pi^\perp$, and the vector subbundle $\ker T\widetilde{\pi} \subset T(\pi^*T^*X)$, defining the foliation $\mathscr{F}_{\widetilde{\pi}}$, whose leaves are the connected components of the fibers of $\widetilde{\pi}$. Therefore, the leaves of the two foliations are mapped onto each other by Ξ.

(iv) By (iii), the smooth map $\widetilde{\pi} \circ \Xi$ is constant on the leaves of the foliation $\mathscr{N}_\pi^\perp$ and thus it drops to a map $\Phi : N_\pi/\mathscr{N}_\pi^\perp \to T^*X$ uniquely characterized by the relation $\Phi \circ \rho = \widetilde{\pi} \circ \Xi$, i.e., $\Phi(\rho(\alpha_b)) := \widetilde{\pi}(\Xi(\alpha_b)) = \beta_{\pi(b)}$.

We prove that Φ is a local diffeomorphism by working in the local coordinates considered earlier. If $\alpha_b \in N_\pi$, $\alpha_b = (x^1, \ldots, x^n, a^1, \ldots, a^k, p_1, \ldots, p_n, 0, \ldots, 0)$, then $\rho(\alpha_b) = (x^1, \ldots, x^n, p_1, \ldots, p_n)$. From (7) and the definition of Ξ, it follows that $\Phi(x^1, \ldots, x^n, p_1, \ldots, p_n) = (x^1, \ldots, x^n, p_1, \ldots, p_n)$, i.e., Φ is the identity map in these charts. Hence Φ is a local diffeomorphism.

Let $\omega_X \in \Omega^2(T^*X)$ and $\omega_B \in \Omega^2(T^*B)$ be the canonical symplectic forms. Then $\Phi^*\omega_X = \omega_\pi$ if and only if $\iota^*\omega_B = \rho^*\omega_\pi = \rho^*\Phi^*\omega_X = \Xi^*\widetilde{\pi}^*\omega_X$ by the definition of the reduced symplectic form ω_π and of the map Φ. The identity $\iota^*\omega_B = \Xi^*\widetilde{\pi}^*\omega_X$ is proved in the local coordinates considered above, in which Ξ is the identity and $\widetilde{\pi}(x^1, \ldots, x^n, a^1, \ldots, a^k, p_1, \ldots, p_n) = (x^1, \ldots, x^n, p_1, \ldots, p_n)$. Therefore,

$$\Xi^*\widetilde{\pi}^*(dx^1 \wedge dp_1 + \cdots + dx^n \wedge dp_n) = dx^1 \wedge dp_1 + \cdots + dx^n \wedge dp_n.$$

On the other hand, $\iota^*\omega_B = \iota^*(dx^1 \wedge dp_1 + \cdots + dx^n \wedge dp_n + da^1 \wedge d\alpha_1 \wedge + \cdots + da^k \wedge d\alpha_k) = dx^1 \wedge dp_1 + \cdots + dx^n \wedge dp_n$, which proves the required identity. Thus, Φ is a symplectic map.

(v) Since $\Phi \circ \rho = \widetilde{\pi} \circ \Xi$ and ρ is surjective, Φ is onto if and only if $\widetilde{\pi}$ is onto, because Ξ is bijective by (i). Since $\widetilde{\pi} : \pi^*T^*X \to T^*X$ is surjective when restricted to every fiber of the vector bundle $\pi^*T^*X \to B$, it follows that $\widetilde{\pi}$ is surjective if and only if π is surjective.

We now study injectivity of ϕ. We first prove that the fiber $\pi^{-1}(x)$, $x \in \pi(B) \subset X$, of π is connected if and only if the fiber $\widetilde{\pi}^{-1}(\beta_x^0)$, $\beta_x^0 \in T_x^*X$ of $\widetilde{\pi}$ is connected. On one hand, the restriction of the smooth map $\pi^*T^*X \ni \left(b, \beta_{\pi(b)}\right) \mapsto b \in B$ to the submanifold $\widetilde{\pi}^{-1}(\beta_x^0)$ of π^*T^*X is a bijective smooth map onto the submanifold $\pi^{-1}(x)$ of B. On the other hand, choose a 1-form β on X such that $\beta(x) = \beta_x^0$ and define the smooth map $B \ni b \mapsto (b, \beta(\pi(b))) \in \pi^*T^*X$. The restriction of this smooth map to the submanifold $\pi^{-1}(x)$ of B maps onto the submanifold $\widetilde{\pi}^{-1}(\beta_x^0)$ of π^*T^*X. These two maps are clearly inverses of each other. Thus, the fibers $\pi^{-1}(x)$

and $\widetilde{\pi}^{-1}(\beta_x^0)$ are diffeomorphic. In particular, $\pi^{-1}(x)$ is connected if and only if $\widetilde{\pi}^{-1}(\beta_x^0)$ is connected for any $x \in \pi(B)$.

We assume now that the fibers of $\widetilde{\pi}$ are connected. Take two classes $[\alpha_b], [\alpha'_{b'}] \in N_\pi/\mathcal{N}_\pi^\perp$ such that $\Phi([\alpha_b]) = \Phi([\alpha'_{b'}])$. The identity $\Phi \circ \rho = \widetilde{\pi} \circ \Xi$ implies that $\widetilde{\pi}(\Xi(\alpha_b)) = \widetilde{\pi}(\Xi(\alpha'_{b'}))$, that is, $\Xi(\alpha_b)$ and $\Xi(\alpha'_{b'})$ are in the same fiber of $\widetilde{\pi}$, which is connected. By (iii), its image by the diffeomorphism Ξ coincides with a leaf of the foliation $\mathcal{N}_\pi^\perp$ and hence α_b and $\alpha'_{b'}$ lie on the same leaf, which means that $[\alpha_b] = [\alpha'_{b'}]$.

Conversely, assume that Φ is injective. Take any two points $\Xi(\alpha_b)$ and $\Xi(\alpha'_{b'})$ in the same fiber of $\widetilde{\pi}$, i.e., $\widetilde{\pi}(\Xi(\alpha_b)) = \widetilde{\pi}(\Xi(\alpha'_{b'}))$, or, equivalently, $\Phi([\alpha_b]) = \Phi([\alpha'_{b'}])$. As Φ injective, we have $[\alpha_b] = [\alpha'_{b'}]$, that is, α_b and $\alpha'_{b'}$, are in the same (connected) leaf of $\mathcal{N}_\pi^\perp$, which is equivalent, by (iii), to $\Xi(\alpha_b)$ and $\Xi(\alpha'_{b'})$ being in the same connected component of the fiber of $\widetilde{\pi}$. Thus, the fibers of $\widetilde{\pi}$, are necessarily connected.

(vi) This is a direct consequence of (iv) and (v). □

3 Morse Families and Transverse Intersections

Let $b \in B$ and $\iota_{\pi^{-1}(\pi(b))} : \pi^{-1}(\pi(b)) \hookrightarrow B$ be the inclusion. For a smooth function $S : B \to \mathbb{R}$ and $b' \in \pi^{-1}(\pi(b))$, denote by $\mathbf{d}^v S(b') := \mathbf{d}S(b')|_{\ker T_{b'}\pi} = \mathbf{d}\left(S \circ \iota_{\pi^{-1}(\pi(b))}\right) : T_{b'}\pi^{-1}(\pi(b)) = \ker T_{b'}\pi \to \mathbb{R}$ the *vertical derivative* at any $b' \in B$. Let

$$\Sigma_S := \{b \in B \mid \mathbf{d}^v S(b) = 0\} \tag{8}$$

be the set of critical points with respect to the projection π. Locally, this states that Σ_S is characterized by points in B with coordinates $(x^1, \ldots, x^n, a^1, \ldots, a^k)$ for which

$$\frac{\partial S}{\partial a^1} = 0, \ldots, \frac{\partial S}{\partial a^k} = 0.$$

Proposition 2 *We have*

$$\mathbf{d}S(B) \cap N_\pi = \mathbf{d}S(\Sigma_S). \tag{9}$$

Proof Indeed, $\mathbf{d}S(b) \in N_\pi = (\ker T\pi)^\circ$ if and only if $\langle \mathbf{d}S(b), v\rangle = 0$, for all $v \in \ker T\pi$, which is equivalent to $b \in \Sigma_S$ by (8).

Definition 1 A function $S : B \to \mathbb{R}$ such that the graph $\mathbf{d}S(B) \subset T^*B$ intersects N_π transversally (i.e., $T_{\mathbf{d}S(b)}\mathbf{d}S(B) + T_{\mathbf{d}S(b)}N_\pi = T_{\mathbf{d}S(b)}(T^*B)$ for any $b \in \Sigma_S$, denoted $\mathbf{d}S(B) \pitchfork N_\pi$) is called a *Morse family*.

Proposition 3 *Given a fibered local system of coordinates* $(x^1, \ldots x^n, a^1, \ldots, a^k)$ *on* B *(i.e., a chart on* B *in which* π *is the projection* $(x^1, \ldots x^n, a^1, \ldots, a^k) \mapsto (x^1, \ldots x^n)$*),* $S : B \to X$ *is a Morse family if and only if the* $k \times (n+k)$*-matrix*

$$\left(\frac{\partial^2 S}{\partial a^i \partial a^j}(b) \quad \frac{\partial^2 S}{\partial a^i \partial x^l}(b)\right) \tag{10}$$

has rank k at every point $b \in \Sigma_S$.

Proof We express the transversality condition $T_{\alpha_b}\mathbf{d}S(B) + T_{\alpha_b}N_\pi = T_{\alpha_b}(T^*B)$ for any $\alpha_b \in \mathbf{d}S(B) \cap N_\pi$ in these local coordinates. Recall that with respect to the standard induced cotangent bundle coordinate system

$$(x^1, \ldots, x^n, a^1, \ldots, a^k, p_1, \ldots, p_n, \alpha_1, \ldots, \alpha_k) \tag{11}$$

on T^*B induced by a fibered system on B, the expression of N_π is $\alpha_1 = \cdots = \alpha_k = 0$ (see (1)). Thus,

$$T_{\alpha_b}N_\pi = \text{span}\left\{\frac{\partial}{\partial x^1}, \ldots, \frac{\partial}{\partial x^n}, \frac{\partial}{\partial a^1}, \ldots, \frac{\partial}{\partial a^k}, \frac{\partial}{\partial p_1}, \ldots, \frac{\partial}{\partial p_n}\right\}.$$

On the other hand, the tangent space to the graph $\mathbf{d}S(B)$ is generated by the vectors

$$\frac{\partial}{\partial x^i} + \frac{\partial^2 S}{\partial x^i \partial x^j}\frac{\partial}{\partial p_j} + \frac{\partial^2 S}{\partial x^i \partial a^r}\frac{\partial}{\partial \alpha_r}, \quad i = 1, \ldots, n, \tag{12}$$

$$\frac{\partial}{\partial a^l} + \frac{\partial^2 S}{\partial a^l \partial x^j}\frac{\partial}{\partial p_j} + \frac{\partial^2 S}{\partial a^l \partial a^r}\frac{\partial}{\partial \alpha_r}, \quad l = 1, \ldots, k. \tag{13}$$

Therefore, for any $\alpha_b \in \mathbf{d}S(B) \cap N_\pi$,

$$T_{\alpha_b}N_\pi + T_{\alpha_b}\mathbf{d}S(B) = \text{span}\left\{\frac{\partial}{\partial x^i}, \frac{\partial}{\partial a^l}, \frac{\partial}{\partial p_i}, \frac{\partial^2 S}{\partial x^i \partial a^r}\frac{\partial}{\partial \alpha_r}, \frac{\partial^2 S}{\partial a^l \partial a^r}\frac{\partial}{\partial \alpha_r}\right\}_{i=1,\ldots,n;\ l=1,\ldots,k},$$

which shows that $T_{\alpha_b}\text{im}(\mathbf{d}S) + T_{\alpha_b}N_\pi = T_{\alpha_b}(T^*B)$ if and only if the coefficient matrix in the statement has maximal rank k. □

Corollary 1 *If S is a Morse family, then the set $\mathbf{d}S(B) \cap N_\pi$ is a submanifold of N_π and $T_{\alpha_b}(\mathbf{d}S(B) \cap N_\pi) = T_{\alpha_b}\mathbf{d}S(B) \cap T_{\alpha_b}N_\pi$, for any $\alpha_b \in \mathbf{d}S(B) \cap N_\pi$. In addition, $\dim(\mathbf{d}S(B) \cap N_\pi) = n$.*

Proof Standard intersection theory (see, e.g., [2, Corollary 3.5.13]) guarantees the first statement. Thus, by linear algebra, $\dim(\mathbf{d}S(B) \cap N_\pi) = \dim \mathbf{d}S(B) + \dim N_\pi - \dim T^*B = (n + k) - (2n + k) - 2(n + k) = n$. □

Theorem 1 *If S is a Morse family and $\rho : N_\pi \to N_\pi/\mathcal{N}_\pi^\perp$ is the projection, then the restriction $\rho|_{\mathbf{d}S(B)\cap N_\pi} : \mathbf{d}S(B) \cap N_\pi \to N_\pi/\mathcal{N}_\pi^\perp$ is an immersion.*

Proof We need to prove that $\ker T_{\alpha_b}(\rho|_{\mathbf{d}S(B)\cap N_\pi}) = \{0\}$, for any $\alpha_b \in \mathbf{d}S(B) \cap N_\pi$. This is equivalent to checking that $T_{\alpha_b}(\mathbf{d}S(B) \cap N_\pi) \cap \ker T_{\alpha_b}\rho = \{0\}$. Note that

$\ker T_{\alpha_b}\rho$ is just the tangent space to the fiber of the foliation $\mathscr{N}_\pi^\perp$ at α_b. In the coordinate system (11), by (3), every tangent vector to the fibers has the expression

$$X = A^l \frac{\partial}{\partial a^l}, \quad X^l \in \mathbb{R}. \tag{14}$$

Suppose $X \in T_{\alpha_b}(\mathbf{d}S(B) \cap N_\pi)$ so, in particular, $X \in T_{\alpha_b}\mathbf{d}S(B)$ and hence X must be a linear combination of the vectors in (12) and (13). Any linear combination having vectors from (12) necessarily contains a linear combination of the vectors $\{\partial/\partial x^i\}_{i=1,\ldots,n}$. The form of the vector (14) precludes this and hence the vector X can only be a linear combination of vectors in (13), i.e.,

$$X = A^l \left(\frac{\partial}{\partial a^l} + \frac{\partial^2 S}{\partial a^l \partial x^j} \frac{\partial}{\partial p_j} + \frac{\partial^2 S}{\partial a^l \partial a^r} \frac{\partial}{\partial \alpha_r} \right)$$

with the condition

$$A^l \frac{\partial^2 S}{\partial a^l \partial x^j} = 0, \qquad A^l \frac{\partial^2 S}{\partial a^l \partial a^r} = 0, \qquad \forall j = 1, \ldots n, \quad \forall l, r = 1, \ldots, k.$$

In matrix form, these conditions are expressed as

$$(A^l) \left(\frac{\partial^2 S}{\partial a^l \partial x^j} \quad \frac{\partial^2 S}{\partial a^l \partial a^r} \right) = \underbrace{(0, \ldots, 0)}_{n+k}.$$

This is only possible if $A^l = 0$, for all $l = 1, \ldots, k$, since the matrix of this linear system is (10) which has maximal rank k by Proposition 3. □

Corollary 2 *The set $\rho(\mathbf{d}S(B) \cap N_\pi)$ is "manifold with self intersections". If we denote by $\rho(\mathbf{d}S(B) \cap N_\pi)_0$ the subset where it is a manifold, then it is a Lagrangian submanifold of $N_\pi / \mathscr{N}_\pi^\perp$.*

Proof The symplectic form ω_π on $N_\pi / \mathscr{N}_\pi^\perp$ is uniquely characterized by the property $\rho^* \omega_\pi = \iota^* \omega_B$, where $\iota : N_\pi \hookrightarrow T^*B$ is the inclusion and ω_B is the canonical symplectic form on T^*B. For any two tangent vectors $U_1, U_2 \in T_x \rho(\mathbf{d}S(B) \cap N_\pi)_0$, $x \in \rho(\mathbf{d}S(B) \cap N_\pi)_0$, there are vectors $X_1, X_2 \in T_{\alpha_b}(\mathbf{d}S(B) \cap N_\pi)$, $\rho(\alpha_b) = x$, such that $T_{\alpha_b}\rho(X_1) = U_1$ and $T_{\alpha_b}\rho(X_2) = U_2$. Then

$$\omega_\pi(x)(U_1, U_2) = \left(\rho^* \omega_\pi\right)(\alpha_b)(X_1, X_2) = \omega_B(\alpha_b)(T_{\alpha_b}\iota(X_1), T_{\alpha_b}\iota(X_2)) = 0,$$

because $X_1, X_2 \in T_{\alpha_b}(\mathbf{d}S(B))$ and $\mathbf{d}S(B)$ is a Lagrangian submanifold of T^*B. By Theorem 1, $\dim\left(\rho(\mathbf{d}S(B) \cap N_\pi)_0\right) = \dim\left(\mathbf{d}S(B) \cap N_\pi\right) = n$ (see Corollary 1). Since $\dim\left(N_\pi / \mathscr{N}_\pi^\perp\right) = 2n$ by Proposition 1(iv), this proves that $\rho(\mathbf{d}S(B) \cap N_\pi)_0$ is a Lagrangian submanifold of $N_\pi / \mathscr{N}_\pi^\perp$. □

Example 1 Let $B := \mathbb{R}^2$, $X := \mathbb{R}$, and $\pi : \mathbb{R}^2 \ni (x, a) \mapsto x \in \mathbb{R}$. As in the general theory, coordinates on $T^*B = T^*\mathbb{R}^2 = \mathbb{R}^4$ are denoted by (x, a, p, α) and on $T^*X = T^*\mathbb{R} = \mathbb{R}^2$ by (x, p). The conormal bundle to the fibers is

$$N_\pi = \{(x, a, p, 0) \mid x, a, p \in \mathbb{R}\} \subset T^*\mathbb{R}^2.$$

We regard N_π as Euclidean space $\mathbb{R}^3$, which enables us to describe the objects of the general theory, for this case, concretely.

Define the function $S : \mathbb{R}^2 \to \mathbb{R}$ by

$$S(x, a) := \frac{a^3}{3} + a(x^2 - 1).$$

Since

$$\Sigma_S = \left\{ (x, a) \in \mathbb{R}^2 \;\middle|\; \frac{\partial S}{\partial a} = a^2 + x^2 - 1 = 0 \right\}$$

and the matrix

$$\left(\frac{\partial^2 S}{\partial x \partial a} \; \frac{\partial^2 S}{\partial a^2} \right) = (2a \; 2x)$$

never vanishes on Σ_S, it follows by Proposition 3 that S is a Morse family. Therefore, the set

$$\mathbf{d}S(\mathbb{R}^2) \cap N_\pi = \{(x, a, 2ax, 0) \mid x^2 + a^2 = 1\} \subset N_\pi = \mathbb{R}^3$$

is a one-dimensional manifold (see Corollary 1). Since π is surjective with connected fibers, by Proposition 1(vi), $N_\pi / \mathcal{N}_\pi^\perp = T^*\mathbb{R} = \mathbb{R}^2$, as symplectic manifolds. Of course, in this special case, this can be shown directly. In fact, in agreement with the general formula (5), the projection $\rho : N_\pi \to N_\pi / \mathcal{N}_\pi^\perp$ is just $\rho(x, a, p, 0) = (x, p)$. Therefore

$$\rho(\mathbf{d}S(\mathbb{R}^2) \cap N_\pi) = \left\{ (x, 2xa) \mid x^2 + a^2 = 1 \right\} = \left\{ \left(x, \pm 2x\sqrt{1 - x^2} \right) \;\middle|\; x \in [-1, 1] \right\}.$$

This is the Bernoulli Lemniscate which clearly has a self-intersection. Away from this self-intersection point, this is a one-dimensional submanifold of $\mathbb{R}^2$ and hence clearly Lagrangian, in agreement with Corollary 2.

4 The Converse: Construction of a Morse Family for a Lagrangian Submanifold

A reasonable converse to Corollary 2 is Theorem 2 below. We use the following notation. If $\varepsilon : A \to B$ is a locally trivial fiber bundle and $M \subset B$ is a submanifold, $\varepsilon_M : A_M \to M$ denotes the restriction of the fiber bundle ε obtained by shrinking the base B to M. For any manifold X, denote by $\pi_X : T^*X \to X$ the cotangent bundle projection.

In the proof of the theorem below, we need a classical result of Weinstein, sometimes called the *Lagrangian Tubular Neighborhood Theorem* or the *Relative Darboux Theorem* (see [5, Theorem 7.1], [6, Lecture 5], [1, Theorem 5.3.18]). We need a general formulation appropriate for our purposes, as stated and proved with all details in [3, Theorem 31.20]. *Let (P, ω) be a symplectic manifold and $L \subset P$ a Lagrangian submanifold. Then there exist an open neighborhood U of L in P, a tubular neighborhood V of the zero section in T^*L, and a symplectic diffeomorphism $\varphi : (U, \omega|_U) \to (V, \omega_L|_V)$ such that $\varphi(x) = 0_x$ for all $x \in L$.* The proof of the theorem starts by considering a Lagrangian complementary vector subbundle E of TL in $(TP)_L$ which ultimately provides the neighborhood U of L in P by constructing a tubular neighborhood of the zero section in E. If such a complement is given, the theorem just cited has an important refinement. *Suppose that $E \to L$ is a given Lagrangian complement to $TL \to L$ in $(TP)_L$. Then the symplectic diffeomorphism φ can be chosen such that $T_x\varphi(E_x) = \ker T_{0_x}\pi_L \subset T_{0_x}(T^*L)$ for all $x \in L$, where $\pi_L : T^*L \to L$ is the cotangent bundle projection.* In other words, $T_x\varphi$ maps the fiber E_x to the vertical vectors at $0_x \in T^*L$ in $T(T^*L)$.

Theorem 2 *Let $L \subset T^*X$ be a Lagrangian submanifold such that*

- *the pull back of the Liouville 1-form $\theta_X \in \Omega^1(T^*X)$ to L is exact, and*
- *there is a Lagrangian subbundle $\Lambda \subset T(T^*X)_L$ which is transversal to both $(\ker T\pi_X)_L$ and TL in $T(T^*X)$.*

Then there is a locally trivial fiber bundle $\pi : B \to X$ and a Morse family $S : B \to \mathbb{R}$ such that $L = (\Phi \circ \rho)(\mathbf{d}S(B) \cap N_\pi)$ (see diagram (6)).

Proof By hypothesis, $\Lambda \to L$ and $TL \to L$ are complementary Lagrangian subbundles. Thus, by the Relative Darboux Theorem cited above, there are open neighborhoods U of L in T^*X and V of the zero section in T^*L, and a symplectic diffeomorphism $f : U \to V$, such that $T_{\alpha_x}f(\Lambda_{\alpha_x}) = \ker T_{0_{\alpha_x}}\pi_L \subset T_{0_{\alpha_x}}(T^*L)$ for all $\alpha_x \in L$. Without loss of generality, we can assume that the fibers $V \cap T_{0_{\alpha_x}}(T^*L)$ are contractible. Since these are fibers of a locally trivial bundle, they form a foliation of V. Therefore, the collection $\left\{f^{-1}\left(V \cap T_{0_{\alpha_x}}(T^*L)\right) \mid \alpha_x \in L\right\}$ forms a foliation on U, the tangent space to every leaf being Λ_{α_x}; these leaves are simply connected, by construction.

Let $\iota : L \hookrightarrow T^*X$ be the inclusion. The pull back of the canonical Liouville 1-form $\theta_L \in \Omega^1(T^*L)$ to the zero section vanishes. Since $f \circ \iota : L \to T^*L$ is the inclusion of the zero section in its cotangent bundle, we conclude that $\iota^*(\theta_X - f^*\theta_L) = \iota^*\theta_X$ is an exact 1-form on L, by hypothesis. As the de Rham cohomology of U, V, and L are isomorphic, it follows that the closed 1-form $\theta_X - f^*\theta_L \in \Omega^1(U)$ is exact. Therefore, there is a smooth function $S : U \to \mathbb{R}$ such that $\mathbf{d}S = \theta_X - f^*\theta_L$.

The vector subbundle $\ker T\pi_X$ is integrable and its leaves are the fibers T_x^*L. Since the vector subbundles $\Lambda \to L$ and $(\ker T\pi_X)_L \to L$ are transversal along L, it follows that their leaves are also transversal at every point of L. Therefore, there is an open neighborhood $B \subset U$ of L, which we choose to have contractible fibers, such that the leaves of these two foliations are transversal at every point of B. Let $\pi : B \to X$ be the restriction of the cotangent bundle projection $\pi_X : T^*X \to X$ to B. Thus $\pi : B \to X$ is a locally trivial fiber bundle.

We show that $\Sigma_S = L$. Let $\beta_x \in B$ and $V_{\beta_x} \in \ker T_{\beta_x}\pi$ be arbitrary. Then $\beta_x \in \Sigma_S$ if and only if $\mathbf{d}S(\beta_x)|_{\ker T_{\beta_x}\pi} = 0$, i.e.,

$$0 = \mathbf{d}S(\beta_x)\left(V_{\beta_x}\right) = \left(\theta_X - f^*\theta_L\right)(\beta_x)\left(V_{\beta_x}\right)$$
$$= \left\langle \beta_x, T_{\beta_x}\pi\left(V_{\beta_x}\right)\right\rangle - \left\langle f(\beta_x), T_{f(\beta_x)}\pi_L T_{\beta_x} f\left(V_{\beta_x}\right)\right\rangle.$$

The first term vanishes since $V_{\beta_x} \in \ker T_{\beta_x}\pi$. The point β_x belongs to a unique leaf of the foliation $\left\{f^{-1}\left(V \cap T_{0_{\alpha_x}}(T^*L)\right) \mid \alpha_x \in L\right\}$, i.e., there is a unique $\alpha_{x_0} \in L$ such that $f(\beta_x) \in V \cap T_{0_{\alpha_{x_0}}}(T^*L)$. This defines a smooth map $g : V \to L$, namely $g(\beta_x)$ is the unique point of intersection of the leaf containing β_x and L. Therefore, $\pi_L \circ f = g$ and the identity above becomes $0 = \left\langle f(\beta_x), T_{\beta_x} g\left(V_{\beta_x}\right)\right\rangle$ for all $V_{\beta_x} \in \ker T_{\beta_x}\pi$. However, $T_{\beta_x} g\left(V_{\beta_x}\right) \neq 0$ for any $V_{\beta_x} \neq 0_{\beta_x}$. Indeed, $T_{\beta_x} g\left(V_{\beta_x}\right) = 0$ if and only if $V_{\beta_x} \in \ker T_{\beta_x} g$ which is the tangent space to the leaf of the foliation $\left\{f^{-1}\left(V \cap T_{0_{\alpha_x}}(T^*L)\right) \mid \alpha_x \in L\right\}$ at β_x. However, this vector space is complementary to $\ker T_{\beta_x}\pi_X$, since the leaf is transversal to $T^*_x L$. Therefore $V_{\beta_x} \in \ker T_{\beta_x}\pi \cap \ker T_{\beta_x} g = \left\{0_{\beta_x}\right\}$. We conclude, therefore, that $0 = \left\langle f(\beta_x), W_{f(\beta_x)}\right\rangle$ for all $W_{f(\beta_x)} \in T_{f(\beta_x)}(T^*L)$. This is equivalent to $f(\beta_x) = 0_{f(\beta_x)}$, i.e., $\beta_x \in L$ since only points in L are mapped by f to the zero section of T^*L.

Note that $\mathbf{d}S(B) \cap N_\pi = \mathbf{d}S(L) = \mathbf{d}S(\Sigma_S)$. Indeed, if $\beta_x \in B$, then $\mathbf{d}S(\beta_x) \in N_\pi = (\ker T\pi)^\circ$ if and only if $\mathbf{d}S(\beta_x)\left(V_{\beta_x}\right) = 0$ for all $V_{\beta_x} \in \ker T_{\beta_x}\pi$. But this is exactly the condition considered above and we conclude that this is equivalent to $\beta_x \in L$.

We finally check that $L = (\Phi \circ \rho)(\mathbf{d}S(B) \cap N_\pi) = (\tilde{\pi} \circ \Xi)(\mathbf{d}S(\Sigma_S))$ (see diagram (6)). The definition of Ξ from Proposition 1(i) yields for $\alpha_x \in L = \Sigma_S$, $\Xi(\mathbf{d}S(\alpha_x)) = (\alpha_x, \beta_x) \in \pi^*T^*X$, where $\beta_x \in T^*_x X$ is defined by

$$\left\langle \beta_x, T_{\alpha_x}\pi\left(V_{\alpha_x}\right)\right\rangle = \left\langle \mathbf{d}S(\alpha_x), V_{\alpha_x}\right\rangle, \quad \forall V_{\alpha_x} \in T_{\alpha_x} B.$$

Therefore, $(\tilde{\pi} \circ \Xi)(\mathbf{d}S(\alpha_x)) = \tilde{\pi}(\alpha_x, \beta_x) = \beta_x$.

We show that $\beta_x = \alpha_x$. Indeed, since

$$\left\langle f^*\theta_L(\alpha_x), V_{\alpha_x}\right\rangle = \left\langle \theta_L(f(\alpha_x)), T_{\alpha_x} f\left(V_{\alpha_x}\right)\right\rangle = \left\langle f(\alpha_x), T_{f(\alpha_x)}\pi_L T_{f\alpha_x}\left(V_{\alpha_x}\right)\right\rangle = 0$$

because $f(\alpha_x) = 0_{\alpha_x}$, we get

$$\left\langle \mathbf{d}S(\alpha_x), V_{\alpha_x}\right\rangle = \left\langle \theta_X(\alpha_x) - (f^*\theta_L)(\alpha_x), V_{\alpha_x}\right\rangle = \left\langle \theta_X(\alpha_x), V_{\alpha_x}\right\rangle = \left\langle \alpha_x, T_{\alpha_x}\pi_L\left(V_{\alpha_x}\right)\right\rangle,$$

which shows that $\alpha_x = \beta_x$ and hence $(\Phi \circ \rho)(\mathbf{d}S(\alpha_x)) = \alpha_x \in L$. □

Acknowledgments Both authors thank the support of the project Edital 061-2011 of the program Ciências sem Fronteiras (CAPES), Brazil, as well as the Pontificia Universidade Católica do Rio de Janeiro for its hospitality in the visit during which this article was written. TSR was partially supported by NCCR SwissMAP of the Swiss National Science Foundation.

References

1. Abraham, R., Marsden, J.E.: Foundations of Mechanics. 2 revised and enlarged edn. Advanced Book Program. Benjamin/Cummings Publishing Co., Inc., Reading (1978). With the assistance of Tudor Ratiu and Richard Cushman
2. Abraham, R., Marsden, J.E., Ratiu, T.S.: Manifolds, Tensor analysis, and Applications. Applied Mathematical Sciences, vol. 75, 2nd edn. Springer, New York (1988)
3. Michor, P.W.: Topics in Differential Geometry. Graduate Studies in Mathematics, vol. 93. American Mathematical Society, Providence (2008)
4. Marsden, J.E., Misiolek, G., Ortega, J.-P., Perlmutter, M., Ratiu, T.S.: Hamiltonian Reduction by Stages. Lecture Notes in Mathematics, vol. 1913. Springer, Berlin (2007)
5. Weinstein, A.: Symplectic manifolds and their Lagrangian submanifolds. Adv. Math. **6**, 329–346 (1971)
6. Weinstein, A.: Lectures on Symplectic Manifolds, Expository lectures from the CBMS Regional Conference held at the University of North Carolina, 8–12 March (1976); Regional Conference Series in Mathematics, vol. 29. American Mathematical Society, Providence (1977)

Special Primes: Properties and Applications

Raúl Durán Díaz and Luis Hernández Encinas

Dedicamos este trabajo al profesor Jaime Muñoz Masqué, en su 65° aniversario, a quien nos sentimos profundamente agradecidos por su abundante y fructífero magisterio, que ha ejercido con generosidad y llaneza: ha sabido ser para nosotros, a la vez, un buen maestro y un maestro bueno.

Abstract This note presents a short survey on current results about the density and methods to obtain several kinds of special primes, together with primality algorithms.

Keywords Primality · Safe primes · Strong primes

1 Introduction

This work summarizes some of Prof. Muñoz Masqué's contributions in the field of primes numbers. We present the properties and counting functions obtained for certain classes of special primes, remarkable for their use and application to several types of cryptosystems. Some of these results are well known and others are proved

---

We dedicate this work to Prof. Jaime Muñoz Masqué in his 65 birthday. We feel deeply indebted to his abundant and fruitful scientific mastery, practised in plain generosity. He became for us a true leader, brimming over with wisdom and bounty.

R. Durán Díaz (✉)
Departamento de Automática, Edificio Politécnico, Campus Universitario, 28871 Alcalá de Henares, Spain
e-mail: raul.duran@uah.es

L. Hernández Encinas
Instituto de Tecnologías Físicas y de la Información, CSIC, C/ Serrano 144, 28006 Madrid, Spain
e-mail: luis@iec.csic.es

M. Castrillón López et al. (eds.), *Geometry, Algebra and Applications: From Mechanics to Cryptography*, Springer Proceedings in Mathematics & Statistics 161, DOI 10.1007/978-3-319-32085-4_7

or conjectured contributions, which aim at giving an idea about the relative density of these classes of primes in the set of natural numbers.

It is interesting to remark that the study of prime numbers is proceeding at good pace, as it is reported in the interesting survey [9], which reviews the achievements of recent years. Prime numbers and related topics are still today and, presumably will be in the future, an enticing source of deep and curious results, which attracts researchers and often becomes a constant supply of surprising connections to many other branches not only of Mathematics but also of other sciences.

The present paper is organized around the results obtained for safe primes (Sect. 2), generalized safe primes (Sect. 3) and optimal strong primes (Sect. 4). For each one of such groups we detail the definition and main properties, and supply the corresponding counting functions.

In order to keep simple the structure, we omit all proofs, which can be found in the appropriate references.

2 Safe Primes with Order *k*

2.1 *Definition and Elementary Properties*

Definition 1 An odd prime number p is said to be a k-safe prime with signature $(\varepsilon_1, \ldots, \varepsilon_k)$, where $\varepsilon_1, \ldots, \varepsilon_k \in \{+1, -1\}$, if k odd prime numbers exist $p_1, \ldots, p_k$ such that

$$p = 2p_1 + \varepsilon_1, \; p_1 = 2p_2 + \varepsilon_2, \; \ldots \; , \; p_{k-1} = 2p_k + \varepsilon_k.$$

The integer k is termed the *order* of the signature.

The set of all prime numbers is denoted by $\mathbb{P}$. For each $k > 0$, the set of k-safe primes with signature $(\varepsilon_1, \ldots, \varepsilon_k)$ is denoted by $\mathbb{P}(\varepsilon_1, \ldots, \varepsilon_k)$. Unless otherwise specified, k-safe primes will be of signature $\varepsilon_1 = \cdots = \varepsilon_k = +1$ and we will write $\mathbb{P}_k^+$ instead of $\mathbb{P}(\varepsilon_1, \ldots, \varepsilon_k)$. We also write $\mathbb{P}_k^- = \mathbb{P}(\varepsilon_1, \ldots, \varepsilon_k)$, as long as $\varepsilon_1 = \cdots = \varepsilon_k = -1$.

Proposition 1 *If $p > 5$ is a k-safe prime with signature $(\varepsilon_1, \ldots, \varepsilon_k)$, then*

$$p \equiv 2^k + \sum_{h=1}^{k} \varepsilon_h 2^{h-1} \pmod{2^{k+1}}.$$

The proof can be carried out by recurrence on k, observing that if p is a k-safe prime with signature $(\varepsilon_1, \ldots, \varepsilon_k)$, then p_1 is $k-1$-safe prime with signature $(\varepsilon'_1 = \varepsilon_2, \ldots, \varepsilon'_{k-1} = \varepsilon_k)$.

Corollary 1 *The sets of k-safe primes with different signatures are mutually disjoint; i.e.,*

$$\mathbb{P}(\varepsilon_1, \ldots, \varepsilon_k) \cap \mathbb{P}(\varepsilon_1', \ldots, \varepsilon_k') = \emptyset, \text{ iff } (\varepsilon_1, \ldots, \varepsilon_k) \neq (\varepsilon_1', \ldots, \varepsilon_k').$$

Corollary 2 *If the set* $\mathbb{P}(\varepsilon_1, \ldots, \varepsilon_k)$ *is empty, so are all the sets with higher order signatures.*

Indeed, by definition, $p \in \mathbb{P}(\varepsilon_1, \ldots, \varepsilon_k)$ if and only if $p = 2p_1 + \varepsilon_1$, with $p_1 \in \mathbb{P}(\varepsilon_2, \ldots, \varepsilon_k)$. Therefore, if the latter set is empty, so must be the former.

Remark 1 Note that the order of the signs inside the signature is relevant, since if π is a permutation of $\{1, \ldots, k\}$, we have that $\mathbb{P}(\varepsilon_1, \ldots, \varepsilon_k) \neq \mathbb{P}(\varepsilon_{\pi(1)}, \ldots, \varepsilon_{\pi(k)})$, in general. For example, $11 \in \mathbb{P}(+1, -1)$ and $11 \notin \mathbb{P}(-1, +1)$, as follows from Corollary 1, since the signature must be regarded as an ordered system; in other words, it is an an element of $\{+1, -1\}^k$. Hence, two signatures are to be considered equal if and only if $\varepsilon_1 = \varepsilon_1', \ldots, \varepsilon_k = \varepsilon_k'$.

2.2 Alternate Signatures

Definition 2 A signature $(\varepsilon_1, \ldots, \varepsilon_k)$ is alternate as long as indices $1 \leq i < j \leq k$ exist such that $\varepsilon_i \varepsilon_j = -1$.

Proposition 2 *The sets* $\mathbb{P}(\varepsilon_1, \ldots, \varepsilon_k)$ *displaying alternate signatures are classified as follows:*

1. *If* $k = 2$, *then*
$$\mathbb{P}(+1, -1) = \{11\}, \quad \mathbb{P}(-1, +1) = \{13\}.$$
2. *If* $k = 3$, *then* $\mathbb{P}(+1, +1, -1) = \{23\}$ *and any other alternate signature is empty.*
3. *If* $k = 4$, *then* $\mathbb{P}(+1, +1, +1, -1) = \{47\}$ *and any other alternate signature is empty.*
4. *If* $k \geq 5$, *all alternate signatures are empty.*

2.3 Chains of Safe Primes

Definition 3 A chain of safe primes of length k is a sequence of prime numbers $p, p_1, \ldots, p_{k-1}$ such that

$$p = 2p_1 + \varepsilon, \; p_1 = 2p_2 + \varepsilon, \ldots, \; p_{k-2} = 2p_{k-1} + \varepsilon, \quad \varepsilon \in \{-1, +1\}.$$

Remark 2 Observe that a prime p sits in the first place of a chain of length k if and only if $p \in \mathbb{P}_{k-1}^+ \cup \mathbb{P}_{k-1}^-$.

Remark 3 Several classes of prime chains are reported in the relevant literature: for example, Cunningham chains and Shanks chains. Our concept of safe prime chain is mostly related to Cunningham's.

A Cunningham chain (see, for example, [10, A7], [19]) is a sequence of $k \geq 2$ primes $p_1, \ldots, p_k$ such that $p_{i+1} = 2p_i + \varepsilon$, $i = 1, \ldots, k-1$, $\varepsilon \in \{-1, +1\}$ (if $\varepsilon = +1$, the chain is termed a Cunningham chain of *first class*, whereas for $\varepsilon = -1$, the chain is termed of *second class*). Remark that for $k = 2$, a Cunningham chain of length 2 is simply the pair $(q, 2q+1)$. In this case, the integer q is known as a Sophie Germain prime.

The Ref. [8] reports a second class Cunningham chain with 16 prime elements, the first one being 3203000719597029781. The current record is held by a pair of second class chains, with 19 primes each, that have been obtained by Raanan Chermoni and Jarosław Wróblewski in March, 2014, as reported by the web page *Cunningham Chain records* (see [1]), and whose first prime elements are 42008163485623434922152331 and 79910197721667870187016101 respectively.

2.4 Counting Function for k-Safe Primes

We define the counting function for k-safe primes as follows:

$$\pi_k^{\pm}(x) = \#\left\{p \in \mathbb{P}_k^{\pm} : p \leq x\right\}.$$

The next result is presented as Lemma 3 in the Ref. [4].

Lemma 1 *Suppose $f_1, f_2, \ldots, f_s \in \mathbb{Z}[x]$, are distinct irreducible polynomials, with integral coefficients, and positive leading coefficients. Suppose F is their product. Let also $Q_F(n)$ be the number of positive integers $j \in [1, n]$ such that $f_1(j)$, $f_2(j)$, ..., $f_s(j)$ are all primes. Then, for large n we have*

$$Q_F(n) \leq 2^s s! C(F) n (\ln n)^{-s} + o(n(\ln n)^{-s}),$$

where

$$C(F) = \prod_{p \in \mathbb{P}} \left(1 - \frac{1}{p}\right)^{-s} \left(1 - \frac{\omega(p)}{p}\right)$$

the product being extended to all primes, and $\omega(p)$ denotes the number of solutions of the congruence

$$F(X) \equiv 0 \pmod{p}.$$

This lemma gave rise to a conjecture that can be found in [3] asserting that, with the same conditions and notation as in Lemma 1, the following approximation holds:

$$Q_F(n) \sim h_1^{-1} h_2^{-1} \dots h_s^{-1} C(F) \int_2^n (\ln u)^{-s} \mathrm{d}u, \tag{1}$$

where $h_1, h_2, \dots, h_s$ represent the degrees of the polynomials $f_1, \dots, f_s$.

This conjecture is clearly applicable to the case of k-safe primes. Indeed, in order to have that $p \in \mathbb{P}_k^{\pm}$, the next set of conditions must be simultaneously satisfied:

(1) p is a prime.
(2) $p_1 = \frac{1}{2}(p \mp 1)$ is a prime.
(3) $p_2 = \frac{1}{2}(p_1 \mp 1) = \frac{1}{4}(p \mp 3)$ is a prime.
 …
$(k+1)$ $p_k = \frac{1}{2}(p_{k-1} \mp 1) = \frac{1}{2^k}(p \mp (2^k - 1))$ is a prime.

It is not difficult to reformulate the previous conditions in such a way that formula (1) be applicable. Actually, if we define $p_k = q$ and we express the successive values of p_i as a function of q, we obtain the following set of polynomials:

$$f_1(q) = q,\ f_2(q) = 2q \pm 1,\ f_3(q) = 4q \pm 3, \dots,\ f_{k+1}(q) = 2^k q \pm (2^k - 1).$$

It is apparent that $p = 2^k q \pm (2^k - 1)$ will be an element of $\mathbb{P}_k^{\pm}$ if all the polynomials $f_i(q)$ take on a prime value for some q.

Following the definition of Q_F, as stated in Lemma 1, it is not difficult to find that it is related to the counting function for k-safe primes as

$$\pi_k^{\pm}(2^k q \pm (2^k - 1)) = Q_F(q).$$

Hence, we finally arrive at

$$\pi_k^{\pm}(x) \sim \frac{1}{2^k} C(F) \int_{2^{k+1} \pm 2^k - 1}^{x} (\ln \tfrac{1}{2^k}(t \mp (2^k - 1)))^{-k-1} \mathrm{d}t. \tag{2}$$

The next step is computing the value for the constant $C(F)$. According to its definition, we have that for our case,

$$C(F) = \prod_{p \in \mathbb{P}} \left(1 - \frac{1}{p}\right)^{-k-1} \left(1 - \frac{\omega(p)}{p}\right).$$

Now, the turn is for the computation of $\omega(p)$ for the admissible values of p. Let us define the following set:

$$V(k) = \{j : p | 2^j - 1,\quad 1 < j \leq k,\quad 2 < p \leq 2^k - 1\}.$$

The solution can be considered on a case-by-case basis as presented below:

$$\omega(p) = \begin{cases} 1, & \text{for } p = 2 \\ p - \#V(k), & \text{for } 2 < p \leq 2^k - 1 \\ k + 1, & \text{for } p > 2^k - 1. \end{cases}$$

As an example, we consider the case $k = 2$, corresponding to 2-safe primes. It is clear that $V(2) = \{1\}$ and, hence, the value of $\omega(p)$ is:

$$\omega(p) = \begin{cases} 1, & \text{for } p = 2 \\ 2, & \text{for } p = 3 \\ 3, & \text{for } p > 3. \end{cases}$$

Therefore, $C(F) = \frac{9}{2} \prod_{p>3} \frac{p^2(p-3)}{(p-1)^3}$. As a consequence, gluing all the pieces together, the counting function eventually results

$$\pi_2^+(x) \sim \frac{9}{8} \prod_{p>3} \frac{p^2(p-3)}{(p-1)^3} \int_{11}^{x} \left(\ln \frac{t-3}{4}\right)^{-3} dt.$$

3 Generalized Safe Primes

Taking a suggestion found in the reference [12], we relax the requirements in Definition 3 by considering chains with numbers of the form

$$p = 2a_1 p_1 + \varepsilon_1,\ p_1 = 2a_2 p_2 + \varepsilon_2, \ldots, p_{k-1} = 2a_k p_k + \varepsilon_k, \tag{3}$$

for certain positive integers $a_1, \ldots, a_k$, which we assume to be of a size smaller than that of the respective primes $p_1, \ldots, p_k$. More information about this type of primes can be found in [6].

Definition 4 For k-safe primes with a vector $\underline{a} = (a_1, \ldots, a_k)$ and signature $\underline{\varepsilon} = (\varepsilon_1, \ldots, \varepsilon_k)$, the counting function is defined as the number $\pi_{\underline{a}}^{\underline{\varepsilon}}(x)$ of $p \in \mathbb{P}$ such that $p \leq x$, and $p = 2a_1 p_1 + \varepsilon_1$, $p_1 = 2a_2 p_2 + \varepsilon_2, \ldots, p_{k-1} = 2a_k p_k + \varepsilon_k$.

The case of positive signatures has received detailed attention in the literature (see, for example, [11–14]). However, alternate and negative signatures have been hardly considered. For this reason, in what follows we turn our attention to them.

3.1 Negative Signature

To begin with, we deal with the case of negative signatures. We split further this case into other two sub-cases, namely, the one in which all the components in the vector $\underline{a}$ are equal, and the contrary. A deeper study, along with the proofs of all the theorems presented in this section can be found in [7].

3.1.1 Vector $\underline{a}$ with Equal Components

First of all, we present the following result:

Theorem 1 *Let us assume that the conjecture giving rise to Eq. (1) is true. If p_a is the least prime divisor of $2a - 1$, then for the vector with equal components $a_1 = \cdots = a_k = a$, and for $\varepsilon_1 = \cdots = \varepsilon_k = -1$, the density of chains in the Eq. (3) is null for all $k \geq p_a - 1$. In particular, this holds if $2a - 1$ is a prime.*

Theorem 2 *Let $a_1 = \cdots = a_k = a$, $\varepsilon_1 = \cdots = \varepsilon_k = -1$, and p_a be as in Theorem 1, and let $D(a) = \{p \in \mathbb{P} : p|a\}$ be the set of prime divisors of $a \in \mathbb{N}$. Suppose that $k \leq p_a - 2$. Then,*

$$\pi_{\underline{a}}^{\underline{\varepsilon}}(x) \sim \frac{C(F)}{(2a)^k} \int_{l_{a,k}}^{x} \left(\ln \left((2a)^{-k} \left[t + \frac{(2a)^k - 1}{2a - 1} \right] \right) \right)^{-k-1} \mathrm{d}t,$$

where $l_{a,k} = \frac{2(2a)^{k+1} - 3(2a)^k + 1}{2a-1}$,

$$C(F) = \prod_{p \in D(2a-1)} \left(1 - \frac{1}{p}\right)^{-k} \left(1 - \frac{k+1}{p}\right) \cdot \prod_{p \in \{2\} \cup D(a)} \left(1 - \frac{1}{p}\right)^{-k} \cdot \prod_{\substack{p \notin D(2a-1) \\ \gcd(a,p)=1}} \left(1 - \frac{1}{p}\right)^{-k} \left(1 - \frac{\min(k+1, e(2a, p))}{p}\right),$$

where $e(2a, p)$ is the order of $2a$ in $\mathbb{Z}_p^*$.

3.1.2 Vector $\underline{a}$ with Unequal Components

For the case of a vector $\underline{a} = (a_1, \ldots, a_k)$ with unequal components, we have not obtained results so sharp as those just presented. We present in the first place some useful notations.

$$\alpha_{hj} = 2^{h-j} \prod_{l=k-h+1}^{k-j} a_l, \quad 1 \le h \le k, \ 0 \le j \le h, \tag{4}$$

$$\beta_h = \sum_{l=1}^{h-1} \alpha_{hl} + 1, \quad 1 \le h \le k, \tag{5}$$

$$N_p^k = \# \left(\{0\} \cup \left\{ \alpha_{h0}^{-1} \beta_h \pmod p) : \alpha_{h0} \in \mathbb{Z}_p^*, 1 \le h \le k \right\} \right), \tag{6}$$

Theorem 3 *Let p be a prime, and let us consider the notations in formulas (4)–(6). If either an index $1 \le h \le k$ exists such that $\alpha_{h0} \equiv 0 \pmod p$ and $\beta_h \equiv 0 \pmod p$, or $N_p^k = p$, then $\omega(p) = p$, and hence $C(F) = 0$.*

Remark 4 Let $p_{\underline{a}} = \max \bigcup_{1 \le h \le k} D(\alpha_{h0})$. If $p \ge p_{\underline{a}} + 1$, then $\alpha_{h0} \not\equiv 0 \pmod p$ for all $1 \le h \le k$.

Theorem 4 *With the same notation and hypothesis as those used in Theorem 3 and Remark 4, if*

$$\{p \in \mathbb{P} : 2 \le p \le k + 1\} \subseteq \bigcup_{1 \le h \le k} \left(D(\alpha_{h0}) \setminus D(\beta_h) \right),$$

then $\omega(p) < p$ for all $p \in \mathbb{P}$, whence $C(F) > 0$.

3.2 Alternate Signatures

Last, we consider the general case of a prime chain $p, p_1, \ldots, p_{k-1}$, $p = 2a_1 p_1 + \varepsilon_1$, $p_1 = 2a_2 p_2 + \varepsilon_2, \ldots, p_{k-1} = 2a_k p_k + \varepsilon_k$, for certain values $\varepsilon_i \in \{-1, +1\}$, $1 \le i \le k$ and indexes $1 \le u < v \le k$ such that $\varepsilon_u \varepsilon_v = -1$.

If $\beta_h = -\sum_{j=1}^{h-1} \alpha_{hj} \varepsilon_{k-j+1} - \varepsilon_{k-h+1}$ then it is easy to check that Theorems 3 and 4 also hold for the case of alternate signatures.

Remark 5 In the case of alternate signatures, $C(F)$ can be zero for some particular value of the vector $\underline{a}$, even if the latter vector has equal components.

Proposition 3 *If the signature of $\mathbb{P}(\varepsilon_1, \ldots, \varepsilon_k)$ is alternate, then $\omega(3) = 3$.*

4 Optimal Strong Primes

A novel concept introduced by Prof. Muñoz Masqué is the notion of *optimal strong prime*. Remarkably, the authors in [18] recommended the use of strong primes for the factors of an RSA modulus. The aim was to prevent attacks from algorithms such as those of Pollard's and Williams' [16, 20] and their improvements [15, 17], along with cyclic attacks. While it is true that the use of strong primes does not give a perfect guarantee against any type of attack (the elliptic curve factorization method can be successful for certain parameters; the algorithm $\Phi_k(p)$ for $k > 2$ might also be lucky, see [2]), they can give additional security at a modest extra cost.

A detailed study of optimal strong primes, together with the proofs of the theorems can be found in [5].

4.1 Standard Definition of a Strong Prime

Definition 5 An odd prime p is said to be strong if it verifies the following three conditions:

(a) $p - 1$ has a large prime factor r.
(b) $p + 1$ has also a large prime factor s.
(c) $r - 1$ has also a large prime factor t.

Remark 6 Strong primes satisfying the conditions in Definition 5 are also called 3-way strong primes.

4.2 The Notion of Optimal Strong Prime

Let us begin with the following

Proposition 4 *Let p an odd prime verifying:*

(a) $p - 1 = ra$, r being an odd prime.
(b) $p + 1 = sb$, s being an odd prime.
(c) $r - 1 = tc$, t being an odd prime.

The next statement holds:

$$a + b + c = \frac{p-1}{r} + \frac{p+1}{s} + \frac{r-1}{t} \geq 12.$$

The proof can be carried out on a case-by-case basis.

The rest of the cases (namely, if any of the values r, s, or t is even) is covered by the following

Proposition 5 *If p is an odd prime not verifying the hypothesis of Proposition 4, then either p is a Fermat (or Mersenne) prime, or p is such that all odd prime factors of $p - 1$ are Fermat primes.*

Definition 6 We say that a strong prime is optimal if the integers r, s, and t in Proposition 4 are as large as can be; or, equivalently, the value

$$a + b + c = \frac{p-1}{r} + \frac{p+1}{s} + \frac{r-1}{t}$$

is as small as possible.

For a given prime p, satisfying the conditions (a)–(c) in Proposition 4, the sum $a + b + c$ takes its minimum value when r, s, t are selected in such a way that they are the largest prime factors of $p - 1$, $p + 1$, $r - 1$, respectively (even if r, s or t are not odd).

Let us draw our attention to the fact that, from Definition 5, a prime may be considered strong if the values r, s and t are "large". Hence, if we choose them so that they are the largest possible ones, we will obtain a "good" strong prime, hopefully, the best one. This fact justifies the selection that we did for that value of $a + b + c$ in the present definition.

Definition 7 Let $S(n)$ the largest prime factor of the integer n if $n \geq 2$ and $S(1) = 1$. We define the function

$$\sigma : \mathbb{N} \setminus \{1, 2\} \to \mathbb{N}$$

by the formula:

$$\sigma(n) = \frac{n-1}{S(n-1)} + \frac{n+1}{S(n+1)} + \frac{S(n-1)-1}{S(S(n-1)-1)}.$$

4.3 Characterization of Optimal Strong Primes

Theorem 5 *For any prime $p \geq 23$ it holds that $\sigma(p) \geq 12$.*

Corollary 3 *A prime p is an optimal strong prime when it verifies all the conditions in Proposition 5 and $\sigma(p)$ takes on its minimum value, namely, $\sigma(p) = 12$.*

Theorem 6 *A prime $p > 29$ is an optimal strong prime if and only if the following conditions are satisfied:*

(i) $\frac{p-1}{6}$ *is 1-safe.*

(ii) $S(p-1) = \frac{p-1}{6}$.

(iii) $S(p+1) = \frac{p+1}{4}$.

Corollary 4 *Obtaining an optimal strong prime is equivalent to finding an integer t such that the integers*

$$t, 2t + 1, 3t + 2, 12t + 7$$

are all simultaneously odd primes.

4.4 Counting Function for Optimal Strong Primes

Definition 8 We define $\pi_\sigma : [0, +\infty) \to \mathbb{N}$ as the counting function for optimal strong primes, assigning to each real number $x \geq 0$ the number of optimal strong primes p such that $p \leq x$. Otherwise expressed,

$$\pi_\sigma(x) = \#\{p \text{ optimal strong prime} : p \leq x\}.$$

According to Corollary 4, generating an optimal strong prime is equivalent to finding a number t, such that t, $2t + 1$, $3t + 2$, $12t + 7$ are all simultaneously prime. Hence, we can immediately apply the results presented in [3], and already provided in Sect. 2.4. The relevant polynomials for this case are $f_1(x) = x$, $f_2(x) = 2x + 1$, $f_3(x) = 3x + 2$, $f_4(x) = 12x + 7$. Therefore, for this case, we have

$$Q_\sigma(y) \sim C_\sigma \int_2^y \frac{du}{(\ln u)^4}, \tag{7}$$

where

$$C_\sigma = \tfrac{42875}{6144} \prod_{p>7} \frac{p^3(p-4)}{(p-1)^4}.$$

It is apparent that the following relation exists

$$Q_\sigma(y) = \pi_\sigma(12y + 7),$$

according to the appropriate definitions. Carrying out the convenient change of variables, we eventually arrive at

$$\pi_\sigma(x) \sim \frac{1}{12} C_\sigma \int_{31}^x \frac{dv}{(\ln \frac{v-7}{12})^4}. \tag{8}$$

The approximate value for the constant C_σ must be computed via numerical methods. If we define

$$C_\sigma(n) = \prod_{q=p_5}^{p_n} \frac{q^3(q-4)}{(q-1)^4},$$

a good approximation for this constant is

$$C_\sigma = \tfrac{42875}{6144} \cdot \lim_{n\to\infty} C_\sigma(n) = 5{,}53491.$$

Acknowledgments This work has been partially funded by both Ministerio de Economía y Competitividad of Spain under grant number TIN2014-55325-C2-1-R (ProCriCiS), and Comunidad de Madrid, Spain, under grant number S2013/ICE-3095-CM (CIBERDINE).

References

1. Augustin, D.: Cunningham chain records (2009). http://primerecords.dk/Cunningham_Chain_records.htm
2. Bach, E., Shallit, J.: Factoring with cyclotomic polynomials. Math. Comput. **52**, 201–219 (1989)
3. Bateman, P., Horn, R.: A heuristic asymptotic formula concerning the distribution of prime numbers. Math. Comput. **16**(79), 363–367 (1962)
4. Bateman, P., Stemmler, R.: Waring's problem for algebraic number fields and primes of the form $(p^r - 1)/(p^d - 1)$. Ill. J. Math. **6**(1), 142–156 (1962)
5. Durán Díaz, R., Muñoz Masqué, J.: Optimal strong primes. Inf. Process. Lett. **93**(1), 47–52 (2005). doi:10.1016/j.ipl.2004.09.015
6. Durán Díaz, R., Hernández Encinas, L., Muñoz Masqué, J.: Computational aspects in the generation of higher-order safe primes. In: Romansky, R. (ed.) Proceedings of the International Conference on Information Technologies, vol. 2, pp. 33–40. Varna, Bulgaria (2008)
7. Durán Díaz, R., Hernández Encinas, L., Muñoz Masqué, J.: Higher-order safe primes with negative signature: an algorithmic approach. Int. J. Inf. Technol. Secur. **1**, 13–24 (2009)
8. Forbes, T.: Prime clusters and Cunningham chains. Math. Comput. **68**(228), 1739–1747 (1999). doi:10.1090/S0025-5718-99-01117-5
9. Granville, A.: Un buen milenio para los primos. La Gaceta de la RSME **12**(3), 547–556 (2009)
10. Guy, R.K.: Unsolved Problems in Number Theory. Springer, New York (1994)
11. Maurer, U.M.: Fast generation of secure RSA-moduli with almost maximal diversity. In: Advances in Cryptology - EUROCRYPT'89, Lecture Notes in Computer Science, vol. 434, pp. 636–647. Springer, Berlin (1990)
12. Maurer, U.M.: Some number-theoretic conjectures and their relation to the generation of cryptographic primes. In: Mitchell, C. (ed.) Cryptography and Coding II, pp. 173–191. Clarendon Press, Oxford (1992). http://citeseer.ist.psu.edu/maurer92some.html
13. Maurer, U.M.: Fast generation of prime numbers and secure public-key cryptographic parameters. J. Cryptol. **8**(3), 123–155 (1995)
14. Mihailescu, P.: Fast generation of provable primes using search in arithmetic progressions. In: Advances in Cryptology - CRYPTO'94, Lecture Notes in Computer Science, vol. 839, pp. 282–293. Springer, Berlin (1994)
15. Montgomery, P., Silverman, R.: An FFT extension to the $p - 1$ algorithm. Math. Comput. **54**(190), 839–854 (1990)
16. Pollard, J.: Theorems on factorization and primality testing. Math. Proc. Camb. Philos. Soc. **76**, 521–528 (1974). doi:10.1017/S0305004100049252
17. Pomerance, C., Sorenson, J.: Counting the integers factorable via cyclotomic methods. J. Algorithms **19**, 250–265 (1995)
18. Rivest, R., Shamir, A., Adleman, L.: A method for obtaining digital signatures and public-key cryptosystems. Commun. ACM **21**(2), 120–126 (1978)
19. Teske, E., Williams, H.: A note on Shanks's chains of primes. In: Proceedings of Algorithmic Number Theory Seminar, ANTS IV, Lecture Notes in Computer Science, vol. 1838, pp. 563–580. Springer, Berlin (2000)
20. Williams, H.: A $p + 1$ method of factoring. Math. Comput. **39**(159), 225–234 (1982)

Rotation Minimizing Vector Fields and Frames in Riemannian Manifolds

Fernando Etayo

Dedicated to Jaime Muñoz Masqué, with a deep gratitude for his generous assistance in the earlier years of my career, on the occasion of his 65th birthday

Abstract We prove that a normal vector field along a curve in $\mathbb{R}^3$ is rotation minimizing (RM) if and only if it is parallel respect to the normal connection. This allows us to generalize all the results of RM vectors and frames to curves immersed in Riemannian manifolds.

Keywords Rotation minimizing · Riemannian manifold · Normal connection

1 Introduction

In a celebrated paper [3], Bishop introduced what nowadays are called *rotation minimizing vector fields* (RM, for short) over a curve in the Euclidean 3-space. The purpose of this note is to show how they can be defined in Riemannian manifolds. This is an expository paper, as self-contained as possible.

When one considers moving orthonormal frames along a curve, one can take into account two general ideas: defining a frame whose first vector is the tangent vector to the curve and defining a frame whose first vector is the unit position vector of the point of the curve. This allows to consider the other two vectors in the normal plane, having only one degree of freedom, as it is explained, e.g. in [8]. Frames having the unit tangent (resp. the position vector) as one component are known as *adapted* (resp. *directed*) *curve frames*.

F. Etayo (✉)
Department of Mathematics, Statistics and Computation,
University of Cantabria, Avda. de Los Castros, s/n, 39071 Santander, Spain
e-mail: etayof@unican.es

M. Castrillón López et al. (eds.), *Geometry, Algebra and Applications: From Mechanics to Cryptography*, Springer Proceedings in Mathematics & Statistics 161, DOI 10.1007/978-3-319-32085-4_8

The most classical adapted frame is the Frenet moving frame, where the second vector is the normal vector to the curve, and the third one is the binormal vector defined as the cross product of tangent and normal vectors (see, e.g., [6]). Given a curve, one can define a swept surface by sweeping out a profile in planes normal to the curve. As it is pointed out in [8], the Frenet frame may result a poor choice for motion planning or swept surface constructions, since it incurs unnecessary rotation of the basis vectors in the normal plane. The fact that the principal normal vector always points to the center of curvature often yields awkward-looking motions, or unreasonably twisted swept surfaces. Besides, in the points where the curvature vanishes one cannot define the Frenet frame. RM frames, which will be defined below, avoid these drawbacks, and they are widely used in Computer Aided Geometric Design. In fact, the Frenet frame of a curve in $\mathbb{R}^n$ can be defined when the curve is *generic*, in the sense that its first $(n-1)$ derivatives are linear independent (see, e.g., [19, p.45]). The Frenet frame of a curve is uniquely defined, gives geometrical information of the curve, but it is defined only in the points where the curve is generic. RM frames are not uniquely defined, and give geometrical information up to rotation, and are defined everywhere. In the paper [2] the authors prove that RM frames in the Euclidean space are preserved by Möbius transformation, i.e., by conformal transformations, considering the normalization of such a transformation. Obviously, one cannot expect the same result for the Frenet frame, because the image of a generic curve under a conformal transformation may not be a generic curve (e.g., a meridian of the sphere goes to a line by the stereographic projection from the north pole to the equatorial plane).

One can define other adapted frames, taking into account the algebraic properties of the ambient Euclidean space. For instance, in the case of plane curves, one can consider $\mathbb{R}^2 = \mathbb{C}$ and define a moving adapted frame given by $\{\mathbf{t}, i\mathbf{t}\}$, i.e., the tangent vector and this vector multiplied by i. This is an RM frame and it can be defined even in the points where the curvature of the curve vanishes, but it gives less geometrical information that the Frenet frame $\{\mathbf{t}, \mathbf{n}\}$, because $\mathbf{n}$ points to the center of curvature, which is not true in general for $i\mathbf{t}$. In the case of curves in $\mathbb{R}^4 = \mathbb{H}$, one can consider the quaternionic structure and define a quaternionic moving frame $\{\mathbf{t}, i\mathbf{t}, j\mathbf{t}, k\mathbf{t}\}$, which is not an RM frame.

Adapted frames can be defined in the case of a curve immersed in a Riemannian manifold, while directed frames has no sense in this general framework, because one cannot define the position vector in a curved manifold. We shall introduce such definition of an RM vector field over a curve in a Riemannian manifold as a vector field parallel respect to the normal connection. It will be shown that this definition is consistent with that of Bishop for curves in the Euclidean space (Sect. 4), and that remains invariable under isometries (Sect. 5). In Sect. 2 we shall remember the basic definitions about RM vector fields and frames, and in Sect. 3 about curves in Riemannian manifolds.

I have chosen this topic to write about, because it has been one of the many topics Prof. Muñoz Masqué has worked, as one can see in [4, 16]. Some of his results will be quoted below.

2 RM Vector Fields and Frames of a Curve in $\mathbb{R}^3$

Let us remember the basic definitions and properties of RM vector fields and frames in the Euclidean space.

Definition 1 A normal vector field $\mathbf{v} = \mathbf{v}(t)$ over a curve $\gamma = \gamma(t)$ in $\mathbb{R}^3$ is said to be *relatively parallel* or *rotation minimizing* (RM) if the derivative $\mathbf{v}'(t)$ is proportional to $\gamma'(t)$.

Then we have:

Remark 1 (1) In this case the ruled surface $f(t, \lambda) = \gamma(t) + \lambda\mathbf{v}(t)$ is developable, because $[\gamma'(t), \mathbf{v}(t), \mathbf{v}'(t)] = 0$.

(2) If $\mathbf{v}$ is an RM vector field, then $\| \mathbf{v} \|$ is constant. Let $\mathbf{t}$ denote the tangent vector to γ. Then $\mathbf{v}' = \lambda\mathbf{t} \Rightarrow \mathbf{v}' \perp \mathbf{v} \Rightarrow \frac{d}{dt}(\mathbf{v} \cdot \mathbf{v}) = 0$.

Besides, one can easily prove the following results:

Remark 2 (1) Let $\gamma(t) = t$ be the line given by the x-axis in $\mathbb{R}^3$. Then a normal vector field $\mathbf{v}(t)$ over γ is RM respect to γ iff it is constant.

(2) Let $\gamma(t) = (\cos t, \sin t, 0)$ the unit circle in the horizontal plane. Let $(0, 0, h)$ any point in the vertical axis. Let us consider the vector $\mathbf{v}(t)$ joining $x(t)$ and $(0, 0, h)$. Then $\mathbf{v}$ is RM. The developable surface generated is the corresponding cone.

(3) Normal and binormal vector fields, $\mathbf{n}$ and $\mathbf{b}$, of a Frenet moving frame are not RM vector fields in general. Let $\gamma = \gamma(s)$ a curve parametrized respect to the arc-length. Then, the Frenet–Serret formulas say that $\mathbf{n}' = -\kappa\mathbf{t} + \tau\mathbf{b}$, and $\mathbf{b}' = -\tau\mathbf{n}$ where κ and τ denote the curvature and the torsion. For a twisted (non plane) curve, $\tau \neq 0$, the above equations show that $\mathbf{n}$ and $\mathbf{b}$ are nor RM vector fields. This is the case, for instance, of the helix $(a\cos t, a\sin t, bt)$. The normal vector is $(-\sin t, \cos t, 0)$ and the surface generated by the normal lines is the helicoid, which is not developable, because its Gauss curvature does not vanish. For a plane curve, $\mathbf{n}$ and $\mathbf{b}$ are RM vector fields.

Definition 2 Let $\gamma = \gamma(t)$ in $\mathbb{R}^3$ be a curve. An RM *frame*, *parallel frame*, *natural frame*, or *Bishop frame* is a moving orthonormal frame $\{\mathbf{t}(t), \mathbf{u}(t), \mathbf{v}(t)\}$ along γ, where $\mathbf{t}(t)$ is the tangent vector to γ at the point $\gamma(t)$ and $\mathbf{u}, \mathbf{v}$ are RM vector fields.

As in the case of Frenet frames, a normal vector field is enough to define an RM frame:

Remark 3 If $\mathbf{u}$ is a unitary RM vector field along γ, then $\{\mathbf{t}, \mathbf{u}, \mathbf{t} \times \mathbf{u}\}$ is an RM frame along γ.

The Frenet frame of a curve is uniquely defined while there exist many RM frames. The following result summarizes the Frenet–Serret-type equations and the Darboux-type vector in a general setting.

Proposition 1 *Let $\gamma = \gamma(s)$ be a curve parametrized respect to the arc length, and let $\{\mathbf{t}, \mathbf{u}, \mathbf{v}\}$ be a moving frame along a curve γ. Then there exists a unique vector field $\omega = A\mathbf{t} + B\mathbf{u} + C\mathbf{v}$ along γ such that the following equations hold:*

$$\frac{d\mathbf{t}}{ds} = \omega \times \mathbf{t} \quad ; \quad \frac{d\mathbf{u}}{ds} = \omega \times \mathbf{u} \quad ; \quad \frac{d\mathbf{v}}{ds} = \omega \times \mathbf{v}.$$

Moreover, the derivatives of the vectors of the frame in terms of the frame are given in matrix expression by the skew symmetric matrix:

$$\begin{pmatrix} 0 & -C & B \\ C & 0 & -A \\ -B & A & 0 \end{pmatrix}.$$

In particular:

(1) If the moving frame is the Frenet frame $\{\mathbf{t}, \mathbf{n}, \mathbf{b}\}$, then $\omega = \kappa\mathbf{b} + \tau\mathbf{t}$ is the Darboux vector field and the above equations are the Frenet–Serret equations

$$\begin{pmatrix} 0 & \kappa & 0 \\ -\kappa & 0 & \tau \\ 0 & -\tau & 0 \end{pmatrix}.$$

(2) The moving frame is an RM frame iff $\omega \cdot \mathbf{t} = 0$. In this case, then Frenet–Serret-type formulas reduce to:

$$\begin{pmatrix} 0 & -C & B \\ C & 0 & 0 \\ -B & 0 & 0 \end{pmatrix}.$$

The component A of the Darboux vector ω measures the rotation of the frame respect to the tangent direction generated by $\mathbf{t}$. In the case of a Frenet frame, $A = -\tau$, and it vanishes iff the curve is plane, thus showing that the Frenet frame is an RM frame iff the curve is plane. In the case of RM frames, A always vanishes, thus showing why they are called 'rotation minimizing'. For the Darboux vector, see, e.g., [22].

Remark 4 One can obtain an RM frame from the Frenet frame by considering the vectors $\mathbf{u}, \mathbf{v}$ given by:

$$\begin{pmatrix} \mathbf{u} \\ \mathbf{v} \end{pmatrix} = \begin{pmatrix} \cos\theta & \sin\theta \\ -\sin\theta & \cos\theta \end{pmatrix} \begin{pmatrix} \mathbf{n} \\ \mathbf{b} \end{pmatrix}$$

where $\theta = -\int \tau \mathrm{ds}$ cancels unnecessary rotation in normal plane.

On the other hand, there exists a direct relation between Frenet 'curvatures' κ, τ and RM 'curvatures' B, C: For the sake of simplicity let us call $\kappa_1 = -C$ and $\kappa_2 = B$, and then the Frenet formulas for RM frames read as:

$$\begin{pmatrix} 0 & \kappa_1 & \kappa_2 \\ -\kappa_1 & 0 & 0 \\ -\kappa_2 & 0 & 0 \end{pmatrix}$$

which is the usual notation (see, e.g., [14, 21]). With this notation one has:

Proposition 2 ([14, p. 52]) *The following relations hold:*

$$\kappa = \sqrt{\kappa_1^2 + \kappa_2^2} \quad \text{and} \quad \tau = \theta' = \frac{\kappa_1 \kappa_2' - \kappa_1' \kappa_2}{\kappa_1^2 + \kappa_2^2}$$

where $\theta = arg(\kappa_1, \kappa_2) = \arctan \frac{\kappa_2}{\kappa_1}$ *and* θ' *is the derivative of* θ *with respect to the arc length.*

Functions κ_i are known as *natural curvatures*, and play the same rôle that curvature and torsion in the Frenet case. They determine uniquely the curve up to an orthogonal transformation. It is a very remarkable fact that Bishop had introduced RM frames before they were interesting in Computer Aided Geometric Design. Nowadays, RM frames are a very useful tool in some aspects of that discipline. See [7] as a basic reference. Besides, they are used in Physics, in the study of solitons [12], and in other Sciences involving the notion of surface growth from a spine curve, such as in the study of DNA [5] and Biology [15].

The above construction of RM frames in the Euclidean space can be generalized to $\mathbb{R}^n$ as Bishop himself pointed out [3].

3 Curves Immersed in a Riemmanian Manifold

In this section we shall remember the main facts about curves immersed in a Riemannian manifold.

As it is well known, if M is a submanifold of a Riemannian manifold $(\overline{M}, \overline{g})$, then the Levi-Civita connection $\overline{\nabla}$ of $(\overline{M}, \overline{g})$ induces a Levi-Civita connection in $(M, g = \overline{g}|_M)$ and a normal connection $D^\perp : \mathfrak{X}(M) \times \Gamma TM^\perp \to \Gamma TM^\perp$, where $\mathfrak{X}(M)$ denotes the module of vector fields of the submanifold and $\Gamma TM^\perp$ the module of sections of the normal bundle. We shall use the notation of [11, VIII].

The normal connection is defined as follows. First of all, consider a point $p \in M$. Then the tangent space at the manifold $\overline{M}$ can be decomposed as a orthogonal direct sum $T_p(\overline{M}) = T_pM \oplus T_p^\perp M$, where T_pM (resp. $T_p^\perp M$) denotes the tangent (resp. the normal) space to the submanifold. The set $TM = \bigcup_{p \in M} T_pM$ (resp. $T^\perp M = \bigcup_{p \in M} T_p^\perp M$) is a manifold called the tangent bundle (res. the normal bundle) and it has a structure of vector bundle over M, given by the natural projection $\pi : TM \to M$; $\pi(u) = p$ if $u \in T_pM$ (resp. $\pi^{\perp} : TM^{\perp} \to M$; $\pi^{\perp}(v) = p$ if $v \in T_p^{\perp}M$). Let us denote by $\mathfrak{X}(M)$ (resp. $\Gamma TM^{\perp}$) the module of vector fields over M, i.e., the module of sections of $\pi : TM \to M$ (resp. the module of sections of $\pi^{\perp} : TM^{\perp} \to M$).

Then for any $X \in \mathfrak{X}(M)$ and $v \in \Gamma TM^{\perp}$ one has the following decomposition, which will be called the *Weingarten formula*:

$$\overline{\nabla}_X v = -A_v X + D_X^{\perp} v$$

where $-A_vX \in \mathfrak{X}(M)$ and $D_X^{\perp} v \in \Gamma TM^{\perp}$. The *Weingarten operator* A is a $\mathfrak{F}(M)$-bilinear map and the normal connection $D^{\perp}$ is a connection in the normal bundle $T^{\perp}M \to M$.

Moreover, if $v, w \in \Gamma TM^{\perp}$ are two normal vector fields, then

$$\overline{g}(D_X^{\perp} v, w) + \overline{g}(v, D_X^{\perp} w) = X(\overline{g}(v, w))$$

which shows that the normal connection $D^{\perp}$ is metric for the fibre metric in the normal bundle $TM^{\perp}$.

A normal vector field $\mathbf{v}$ is said to be *parallel* respect to $X \in \mathfrak{X}(M)$ if $D_X^{\perp} v = 0$.

Let $\pi : E \to M$ be a vector bundle with a connection D, compatible with a metric on E. The parallel transport induced by D defines an isometry between any two different fibres of $\pi : E \to M$ (see [18] for details). Thus, the norm of a parallel section remains constant and the angle between two parallel sections also remains constant. In the present case of having a submanifold M of a Riemannian manifold $(\overline{M}, \overline{g})$, the vector bundle we are considering is the normal bundle, and the metric is the restriction of $\overline{g}$ to normal vectors.

Many results about curves in Riemannian manifolds have been obtained in the past. We would like to point out that generalizations of Frenet frames have been obtained in [13] and [9], in [16] for the case of spaces of constant curvature, and in [17] for the case of the Minkowski space. Besides in [4] some results about the total curvature of a curve in a Riemannian manifold are also obtained.

The aim of the present note is not to show Frenet frames in Riemannian manifolds, but RM frames. Nevertheless, I would like to comment the very beautiful main results of the paper [16]: (1) a Frenet theorem: two curves in the Euclidean, Spherical or Hyperbolic space are congruent if and only if their $n-1$ curvatures are equal, and (2) the converse: Frenet's theorem holds for curves in a connected Riemannian manifold (M, g) if and only if (M, g) is of constant curvature. Thus, one cannot expect to extend this Theorem to other Riemannian manifolds.

4 RM Vector Fields Over a Curve Immersed in a Riemmanian Manifold

The notion of RM vector field implies a notion of parallel transport. We shall show this carefully.

Let us consider the case where $(\overline{M}, \overline{g})$ is $\mathbb{R}^3$ with the standard product and $(M, g = \overline{g}|_M)$ is a curve γ. Let us denote by $T_{\gamma(t_0)}$ (resp. $T^{\perp}_{\gamma(t_0)}$) the tangent line (resp. the normal plane) to γ in $\gamma(t_0)$. Then we have:

Theorem 1 *A normal vector field* $\mathbf{v}$ *over a curve* γ *immersed in* $\mathbb{R}^3$ *is an RM vector field iff it is parallel respect to the normal connection of* γ.

Proof Let us denote as (x^1, x^2, x^3) the global coordinates in $\mathbb{R}^3$. The curve γ can be expressed as $\gamma(s) = (\gamma^1(s), \gamma^2(s), \gamma^3(s))$, s being the arc-length parameter, and the tangent vector is $\overline{\mathbf{t}} = \gamma'(s) = \frac{\partial}{\partial x^i}\frac{d\gamma^i}{ds}$ (Einstein's convention is assumed).

Let $\mathbf{v}$ be a normal vector field over γ, $\mathbf{v} = \frac{\partial}{\partial x^i}v^i$. The condition of being normal to the curve means that

$$\sum_{i=1}^{3} v^i \frac{d\gamma^i}{ds} = 0.$$

The condition of being $\mathbf{v}$ an RM vector field means that $\mathbf{v}'(t)$ is proportional to $\gamma'(t)$, i.e.

$$\frac{dv^i}{ds} = \lambda(s)\frac{d\gamma^i}{ds}, \qquad \forall i = 1, 2, 3$$

where $\lambda = \lambda(s)$ is a function.

Now, we shall check the value of $D^{\perp}_{\gamma'(s)}\mathbf{v}$. We must prove that $D^{\perp}_{\gamma'(s)}\mathbf{v} = 0$ iff the above equation is satisfied. Let $\overline{\nabla}$ be the Levi-Civita connection of $\mathbb{R}^3$. The all of its Christoffel symbols vanish, and then one has:

$$\overline{\nabla}_{\overline{\mathbf{t}}}\,\mathbf{v} = \overline{\nabla}_{\frac{\partial}{\partial x^i}\frac{d\gamma^i}{ds}}\left(\frac{\partial}{\partial x^j}v^j\right) = \frac{d\gamma^i}{ds}\overline{\nabla}_{\frac{\partial}{\partial x^i}}\left(\frac{\partial}{\partial x^j}v^j\right) = \frac{\partial}{\partial x^j}\frac{d\gamma^i}{ds}\frac{\partial v^j}{\partial x^i}.$$

Applying the chain rule one has:

$$\frac{\partial v^j}{\partial x^i} = \frac{dv^j}{ds}\frac{ds}{dx^i} = \frac{dv^j}{ds}\left(\frac{dx^i}{ds}\right)^{-1} = \frac{dv^j}{ds}\left(\frac{d(x^i\circ\gamma^{-1})}{ds}\right)^{-1} = \frac{dv^j}{ds}\left(\frac{d\gamma^i}{ds}\right)^{-1}.$$

And then,

$$\overline{\nabla}_{\overline{\mathbf{t}}}\,\mathbf{v} = \frac{\partial}{\partial x^j}\frac{d\gamma^i}{ds}\frac{\partial v^j}{\partial x^i} = \frac{\partial}{\partial x^j}\frac{d\gamma^i}{ds}\frac{dv^j}{ds}\left(\frac{d\gamma^i}{ds}\right)^{-1} = \frac{\partial}{\partial x^j}\frac{dv^j}{ds}.$$

Finally, $\mathbf{v}$ is parallel $\Leftrightarrow D^{\perp}_{\gamma'(s)}\mathbf{v} = 0 \Leftrightarrow \overline{\nabla}_{\bar{\mathbf{t}}}\mathbf{v}$ is tangent to $\gamma \Leftrightarrow \frac{dv^i}{ds} = \lambda(s)\ \frac{d\gamma^i}{ds}$, $\forall i = 1, 2, 3 \Leftrightarrow \mathbf{v}$ is an RM vector field, thus finishing the proof. □

The above result is important because it allows to obtain the definition of an RM vector field over a curve immersed in a Riemmannian manifold. Moreover, one easily can deduce the following properties of the above Proposition.

Corollary 1 *With the above notation:*

(1) Given a vector $\mathbf{v}_0 \in T^{\perp}_{\gamma(t_0)}$ *there exists a unique RM vector field* $\mathbf{v}$ *over* γ *such that* $\mathbf{v}(t_0) = v_0$.

(2) If $\mathbf{v}$ *is an RM vector field over* γ *then the norm* $\|\mathbf{v}\|$ *is constant.*

(3) If $\mathbf{v}$ *and* $\mathbf{w}$ *are RM vector fields over* γ *then the angle between* $\mathbf{v}(t)$ *and* $\mathbf{w}(t)$ *is constant.* □

Thus, we can give the following:

Definition 3 Let γ be a curve immersed in a Riemannian manifold $(\overline{M}, \overline{g})$.

(1) A normal vector field $\mathbf{v}$ over γ is said to be an *RM vector field* if it is parallel respect to the normal connection of γ.

(2) A *parallel frame*, *natural frame*, or RM *frame* is a moving orthonormal frame

$$\{\mathbf{t}(t), \mathbf{v}_1(t), \dots, \mathbf{v}_n(t)\}$$

along γ, where $\mathbf{t}(t)$ is the tangent vector to γ at the point $\gamma(t)$ and $\mathbf{v}_i$ are RM vector fields, $\forall i \in 1, \dots, n$.

If one defines an orthonormal frame $\{\mathbf{t}(t_0), \mathbf{v}_1(t_0), \dots, \mathbf{v}_n(t_0)\}$ at a point $\gamma(t_0)$ of a curve γ then by parallel transport it can be extended along γ. Parallel transport is an isometry, this meaning that norms and angles are preserved.

Remark 5 Taking into account the Weingarten formula one can easily check that an RM frame satisfies:

$$\overline{\nabla}_{\mathbf{t}}\, \mathbf{t} \perp \mathbf{t} \quad ; \quad \overline{\nabla}_{\mathbf{t}}\, \mathbf{v}_i \parallel \mathbf{t}$$

where $\overline{\nabla}$ is the Levi-Civita connection of $(\overline{M}, \overline{g})$, thus coinciding our definition with that given in [1]. The first equation is an easy consequence of the well-known identity $\overline{g}(\overline{\nabla}_X Y, Z) + \overline{g}(Y, \overline{\nabla}_X Z) = X(\overline{g}(Y, Z))$, when one consider $X = Y = Z$ vector extensions of $\overrightarrow{t}$, which is a vector field along γ of constant norm equal to 1.

As in the case of the Euclidean space, any parallel moving frame differs from a classical Frenet frame by a point-dependent $SO(n-1)$ rotation acting in the normal space of the curve γ. As the components of a Frenet connection matrix along γ are differential invariants of the curve, then the components of a parallel connection matrix are invariantly defined up to the covariant action of the equivalence group of rigid $SO(n-1)$ rotations. The Frenet-type equations have a similar expression to that given in Proposition 1 (see [13] or [20]).

RM frames in Riemannian manifolds are used in the study of the structure equations for the evolution of a curve embedded in an n-dimensional Riemannian manifold with constant curvature (see, e.g., [13, 20]) or a symmetric Riemannian space (see [1]). They are also used in the study of mathematical models of equilibrium configurations of thin elastic rods (see, e.g., [10] and the references therein).

5 RM Frames and Transformations

We prove that RM vector fields and frames are preserved by isometries. Let $\mu : (\overline{M}, \overline{g}) \to (\overline{M}, \overline{g})$ be an isometry and let $\mu_{*p} : T_p\overline{M} \to T_p\overline{M}$ its differential or tangent map. Then μ_{*p} is a linear isometry respect to $\overline{g}_p$, i.e., $\overline{g}(\mu_*\mathbf{v}, \mu_*\mathbf{w}) = \overline{g}(\mathbf{v}, \mathbf{w})$.

Theorem 2 *Let γ be a curve immersed in a Riemannian manifold $(\overline{M}, \overline{g})$ and let $\mu : (\overline{M}, \overline{g}) \to (\overline{M}, \overline{g})$ be an isometry.*

(1) If $\mathbf{v}$ is an RM vector field over γ, then $\mu_*(\mathbf{v})$ is an RM vector field over $\mu \circ \gamma$.

(2) If $\{\mathbf{t}, \mathbf{v}_1, \dots, \mathbf{v}_n\}$ is an RM frame over γ, then $\{\mu_*(\mathbf{t}), \mu_*(\mathbf{v}_1), \dots, \mu_*(\mathbf{v}_n)\}$ is an RM frame over $\mu \circ \gamma$.

Proof The following claims are well known:

1. If $\mathbf{t}(t_0)$ is the tangent vector of γ at the point $\gamma(t_0)$, then $\mu_*(\mathbf{t}(t_0))$ is the tangent vector of $\mu \circ \gamma$ at the point $(\mu \circ \gamma)(t_0)$.
2. If $\mathbf{v} \in T^{\perp}_{\gamma(t_0)}$, then $\mu_*(\mathbf{v}) \in T^{\perp}_{(\mu\circ\gamma)(t_0)}$, because μ_* is an isometry.
3. $\mu_*(\overline{\nabla}_X Y) = \overline{\nabla}_{\mu_* X} \; \mu_* Y$ (cfr., e.g. [11, p. 161, vol. 1]).

Claims (1) and (2) show that μ_* maps tangent (resp. normal) vectors in tangent (resp. normal vectors).

Now, we can easily prove the theorem.

(1) Let $\mathbf{v}$ be an RM vector field over γ. Then $D^{\perp}_{\mathbf{t}}\mathbf{v} = 0$, where $\mathbf{t}$ denotes the tangent vector to γ. We must prove that $D^{\perp}_{\mu_*\mathbf{t}} \; \mu_*\mathbf{v} = 0$.

We have the tangent and normal decomposition:

$$\overline{\nabla}_{\mu_*\mathbf{t}} \; \mu_*\mathbf{v} = -A_{\mu_*\mathbf{v}} \; \mu_*\mathbf{t} + D^{\perp}_{\mu_*\mathbf{t}} \; \mu_*\mathbf{v}$$

and, on the other hand,

$$\overline{\nabla}_{\mu_*\mathbf{t}} \; \mu_*\mathbf{v} = \mu_*(\overline{\nabla}_{\mathbf{t}} \; \mathbf{v}) = \mu_*(-A_v\mathbf{t} + D^{\perp}_{\mathbf{t}} \; v) = \mu_*(-A_v\mathbf{t})$$

which is tangent to $(\mu \circ \gamma)$, thus proving $D^{\perp}_{\mu_*\mathbf{t}} \; \mu_*\mathbf{v} = 0$.

(2) It's a direct consequence of part (1) and claims 1 and 2 at the beginning of the proof. □

Acknowledgments The author wants to express his gratitude to his colleagues Marco Castrillón, Laureano González-Vega, Bert Jütler and Gema Quintana for the useful talks about the theory of RM vectors and frames.

References

1. Anco, S.C.: Group-invariant soliton equations and bi-Hamiltonian geometric curve flows in Riemannian symmetric spaces. J. Geom. Phys. **58**, 1–37 (2008)
2. Bartoň, M., Jüttler, B., Wang, W.: Construction of rational curves with rational rotation-minimizing frames via Möbius transformations. Mathematical Methods for Curves and Surfaces. Lecture Notes in Computer Science, vol. 5862, pp. 15–25. Springer, Berlin (2010)
3. Bishop, R.L.: There is more than one way to frame a curve. Am. Math. Mon. **82**, 246–251 (1975)
4. Castrillón López, M., Fernández Mateos, V., Muñoz Masqué, J.: Total curvature of curves in Riemannian manifolds. Differ. Geom. Appl. **28**, 140–147 (2010)
5. Clauvelin, N., Olson, W.K., Tobias, I.: Characterization of the geometry and topology of DNA pictured as a discrete collection of atoms. J. Chem. Theory Comput. **8**(3), 1092–1107 (2012)
6. do Carmo, M.: Differential Geometry of Curves and Surfaces. Prentice-Hall, Englewood Cliffs (1976)
7. Farouki, R.T.: Pythagorean-Hodograph Curves: Algebra and Geometry Inseparable. Geometry and Computing, vol. 1. Springer, Berlin (2008)
8. Gianelli, C.: Rational moving frames on polynomial space curves: theory and applications. Ph.D. Thesis, Università degli studi di Firenze, Florence (2009)
9. Gutkin, E.: Curvatures, volumes and norms of derivatives for curves in Riemannian manifolds. J. Geom. Phys. **61**, 2147–2161 (2011)
10. Kawakubo, S.: Kirchhoff elastic rods in five-dimensional space forms whose centerlines are not helices. J. Geom. Phys. **76**, 158–168 (2014)
11. Kobayashi, S., Nomizu, K.: Foundations of Differential Geometry, vol. I and II. Interscience Publishers (a division of Wiley), New York (1963, 1969)
12. Langer, J.: Recursion in curve geometry. New York J. Math. **5**, 25–51 (1999)
13. Marí Beffa, G.: Poisson brackets associated to invariant evolutions of Riemannian curves. Pac. J. Math. **215**(2), 357–380 (2004)
14. McCreary, P.R.: Visualizing Riemann surfaces, Teichmüller spaces, and transformations groups on hyperbolic manifolds using real time interactive computer animator (RTICA) graphics. Ph.D. Thesis, University of Illinois at Urbana-Champaign (1998)
15. Moulton, D.E., Goriely, A.: Surface growth kinematics via local curve evolution. J. Math. Biol. **68**(1–2), 81–108
16. Muñoz Masqué, J., Rodríguez Sánchez, G.: Frenet theorem for spaces of constant curvature. Geometry from the Pacific Rim (Singapore, 1994), 253–259, de Gruyter, Berlin (1997)
17. Özdemir, M., Ergin, A.A.: Parallel frame of non-lightlike curves. Missouri J. Math. Sci. **20**(2), 1–10 (2008)
18. Poor, W.A.: Differential Geometric Structures. McGraw-Hill Book Co., New York (1981)
19. Postnikov, M.: Lectures in Geometry. Semester III. Mir, Moscow (1989)
20. Sanders, J.A., Wang, J.P.: Integrable systems in n-dimensional Riemannian geometry. Mosc. Math. J. **3**(4), 1369–1393 (2003)
21. Singer, D.A.: Lectures on elastic curves and rods. Curvature and Variational Modeling in Physics and Biophysics. In: AIP Conference Proceedings, vol. 1002, pp. 3–32 (2008)
22. Wang, W., Jüttler, B., Zheng, D., Liu, Y.: Computation of rotation minimizing frames. ACM Trans. Graph. (TOG) **27**(1), 1–18 (2008)

Local Anomaly Cancellation and Equivariant Cohomology of Jet Bundles

Roberto Ferreiro Pérez

Dedicated to Jaime Muñoz Masqué on the occasion of his 65th birthday, with my best wishes.

Abstract We study the problem, suggested by Singer in [17], and consisting in determining a notion of "local cohomology" adequate to deal with the problem of locality in those approaches to local anomalies based on the Atiyah–Singer index theorem.

Keywords Local cohomology · Equivariant cohomology · Jet bundle · Anomaly cancellation

1 Introduction

An anomaly appears in a theory when a classical symmetry is broken at the quantum level. As we consider only local anomalies, we can assume that the group $\mathcal{G}$ is connected. Let $\mathcal{L}(\psi, s)$ be a $\mathcal{G}$-invariant Lagrangian density depending on bosonic fields $s \in \Gamma(E)$ and fermionic fields ψ. At the quantum level, the corresponding effective action $W(s)$, defined in terms of the fermionic path integral by $\exp(-W(s)) = \int \mathcal{D}\psi\mathcal{D}\bar{\psi} \exp\left(-\int_M \mathcal{L}(\psi, s)\right)$ could fail to be $\mathcal{G}$-invariant. We define a form $\mathcal{A} \in \Omega^1(\mathrm{Lie}\,\mathcal{G}, \Omega^0(\Gamma(E)))$ by $\mathcal{A} = \delta W$, i.e. $\mathcal{A}(X)(s) = L_X W(s)$ for $X \in \mathrm{Lie}\,\mathcal{G}$, $s \in \Gamma(E)$. Although W is clearly a non-local functional, $\mathcal{A}$ is local in X and s, i.e. we have $\mathcal{A} \in \Omega^1_{\mathrm{loc}}(\mathrm{Lie}\,\mathcal{G}, \Omega^0_{\mathrm{loc}}(\Gamma(E)))$. It is clear that $\mathcal{A}$ satisfies the

R. Ferreiro Pérez (✉)
Facultad de Ciencias Económicas y Empresariales, Departamento de Economa Financiera y Contabilidad I, Universidad Complutense de Madrid, Campus de Somosaguas, 28223 Pozuelo de Alarcón, Spain
e-mail: roferreiro@ccee.ucm.es

M. Castrillón López et al. (eds.), *Geometry, Algebra and Applications: From Mechanics to Cryptography*, Springer Proceedings in Mathematics & Statistics 161, DOI 10.1007/978-3-319-32085-4_9

condition $\delta\mathcal{A} = 0$ (the Wess–Zumino consistency condition). Moreover, if $\mathcal{A} = \delta\Lambda$ for a *local* functional $\Lambda = \int_M \lambda \in \Omega^0_{\mathrm{loc}}(\Gamma(E))$ then we can define a new lagrangian density $\hat{\mathcal{L}} = \mathcal{L} + \lambda$, such that the new effective action $\hat{W}$ is $\mathcal{G}$-invariant, and in that case the anomaly cancels. If $\mathcal{A} \neq \delta\Lambda$ for every $\Lambda \in \Omega^0_{\mathrm{loc}}(\Gamma(E))$ then we say that there exists an anomaly in the theory. Hence the anomaly is measured by the cohomology class of $\mathcal{A}$ in the BRST cohomology $H^1_{\mathrm{loc}}(\mathrm{Lie}\,\mathcal{G}, \Omega^0_{\mathrm{loc}}(\Gamma(E)))$ (e.g. see [4, 6–8, 16]).

Local anomalies also admit a nice geometrical interpretation in terms of the Atiyah–Singer index theorem for families of elliptic operators (see [1, 2, 4, 17]). The first Chern class $c_1\,(\det \mathrm{Ind} D) \in H^2(\Gamma(E)/\mathcal{G})$ of the determinant line bundle $\det \mathrm{Ind} D \to \Gamma(E)/\mathcal{G}$ represents an obstruction for anomaly cancellation. However, the condition $c_1\,(\det \mathrm{Ind} D) = 0$ is a necessary but not a sufficient condition for local anomaly cancellation due to the problem of locality. In [17] (see also [1]) Singer proposes the problem of defining a notion of "local cohomology of $\Gamma(E)/\mathcal{G}$", $H^2_{\mathrm{loc}}(\Gamma(E)/\mathcal{G})$, adequate to study local anomaly cancellation. The principal difficulty is the fact that the expression of the curvature of $\det \mathrm{Ind} D$ itself contains non-local terms (Green operators).

Moreover, we recall (see [2, 5, 15]) that the BRST and index theory approaches are related by means of the transgression map t (see Sect. 2), i.e., we have $[\mathcal{A}] = t(c_1\,(\det \mathrm{Ind} D))$. As t is injective, the condition $c_1\,(\det \mathrm{Ind} D) = 0$ on $H^2(\Gamma(E)/\mathcal{G})$ is equivalent to $[\mathcal{A}] = 0$ on $H^1(\mathrm{Lie}\,\mathcal{G}, \Omega^0(\Gamma(E)))$. However, the condition for local anomaly cancellation is $[\mathcal{A}] = 0$ on the BRST cohomology $H^1_{\mathrm{loc}}(\mathrm{Lie}\,\mathcal{G}, \Omega^0_{\mathrm{loc}}(\Gamma(E)))$. We define $H^\bullet_{\mathrm{loc}}(\Gamma(E)/\mathcal{G})$ in such a way that the preceding condition is equivalent to the vanishing of the class of $c_1\,(\det \mathrm{Ind} D)$ on $H^2_{\mathrm{loc}}(\Gamma(E)/\mathcal{G})$, hence solving Singer's problem.

2 The Transgression Maps

First we recall the definition of equivariant cohomology in the Cartan model (*e.g.* see [3]). We consider a left action of a connected Lie group $\mathcal{G}$ on a manifold $\mathcal{N}$. We have an induced Lie algebra homomorphism $\mathrm{Lie}\,\mathcal{G} \to \mathfrak{X}(\mathcal{N})$, $X \mapsto X_{\mathcal{N}} = \frac{d}{dt}\big|_{t=0} \rho(\exp(-tX))$. We denote by $\mathcal{P}^k(\mathrm{Lie}\,\mathcal{G}, \Omega^r(\mathcal{N}))^{\mathcal{G}}$ the space of degree k $\mathcal{G}$-invariant polynomials on $\mathrm{Lie}\,\mathcal{G}$ with values in $\Omega^r(\mathcal{N})$. We recall that $\alpha \in \mathcal{P}^k(\mathrm{Lie}\,\mathcal{G}, \Omega^r(\mathcal{N}))^{\mathcal{G}}$ if and only if $\alpha(\mathrm{Ad}_g X) = \rho(g^{-1})^*(\alpha(X))\, \forall X \in \mathrm{Lie}\,\mathcal{G}, \forall g \in \mathcal{G}$.

The space of $\mathcal{G}$-equivariant differential q-forms is defined by

$$\Omega^q_{\mathcal{G}}(\mathcal{N}) = \bigoplus_{2k+r=q} (\mathcal{P}^k(\mathrm{Lie}\,\mathcal{G}, \Omega^r(\mathcal{N})))^{\mathcal{G}}. \tag{1}$$

The Cartan differential $d_c\colon \Omega^q_{\mathcal{G}}(\mathcal{N}) \to \Omega^{q+1}_{\mathcal{G}}(\mathcal{N})$, $(d_c\alpha)(X) = d(\alpha(X)) - \iota_{X_{\mathcal{N}}}\alpha(X)$ for $X \in \mathrm{Lie}\,\mathcal{G}$, satisfies $(d_c)^2 = 0$, and the $\mathcal{G}$-equivariant cohomology of $\mathcal{N}$, $H^q_{\mathcal{G}}(\mathcal{N})$, is the cohomology of this complex.

We recall (e.g. see [3]) that if $\mathcal{N} \to \mathcal{N}/\mathcal{G}$ is a principal $\mathcal{G}$-bundle we have the (generalized) Chern–Weil homomorphism ChW: $H^\bullet_{\mathcal{G}}(\mathcal{N}) \to H^\bullet(\mathcal{N}/\mathcal{G})$. If A is an arbitrary connection on $\mathcal{N} \to \mathcal{N}/\mathcal{G}$ with curvature F_A, and $\alpha \in \Omega^q_{\mathcal{G}}(\mathcal{N})$, then we have $\mathrm{ChW}([\alpha]) = [\mathrm{hor}_A(\alpha(F_A))]$, where hor_A is the horizontalization with respect to the connection A. We also use the notation $\underline{\alpha} = \mathrm{ChW}(\alpha)$.

If $\omega \in \Omega^2_{\mathcal{G}}(\mathcal{N})$ is a closed $\mathcal{G}$-equivariant 2-form, then we have $\omega = \omega_0 + \mu$ where $\omega_0 \in \Omega^2(\mathcal{N})$ is closed, and $\mu\colon \mathrm{Lie}\,\mathcal{G} \to C^\infty(\mathcal{N})$, is a $\mathcal{G}$-equivariant moment map for ω_0, i.e., $\iota_{X_\mathcal{N}}\omega_0 = d(\mu(X))$ for $X \in \mathrm{Lie}\,\mathcal{G}$. A direct computation shows that we have the following

Proposition 1 *Assume that $\mathcal{N} \to \mathcal{N}/\mathcal{G}$ is a principal $\mathcal{G}$-bundle, and let $A \in \Omega^1(\mathcal{N}, \mathrm{Lie}\,\mathcal{G})$ be a connection form. If $\omega = \omega_0 + \mu \in \Omega^2_{\mathcal{G}}(\mathcal{N})$ is a closed $\mathcal{G}$-equivariant 2-form and we define $\alpha \in \Omega^1(\mathcal{N})^{\mathcal{G}}$ by $\alpha(X) = \mu(A(X))(x)$ for $X \in T_x\mathcal{N}$, then we have $\mathrm{ChW}_A(\omega) = \omega + d_c\alpha$.*

Corollary 1 *The map* ChW: $H^2_{\mathcal{G}}(\mathcal{N}) \to H^2(\mathcal{N}/\mathcal{G})$ *is an isomorphism.*

Let us assume now that $H^1(\mathcal{N}) = H^2(\mathcal{N}) = 0$. The cohomology of the Lie algebra $\mathrm{Lie}\,\mathcal{G}$ with values in $\Omega^0(\mathcal{N})$ is denoted by $H^\bullet(\mathrm{Lie}\,\mathcal{G}, \Omega^0(\mathcal{N}))$.

Proposition 2 *Let $\omega = \omega_0 + \mu \in \Omega^2_{\mathcal{G}}(\mathcal{N})$ be a closed $\mathcal{G}$-equivariant form. If $\rho \in \Omega^1(\mathcal{N})$ satisfies $\omega_0 = d\rho$, then the map $\tau_\rho \in \Omega^1(\mathrm{Lie}\,\mathcal{G}, \Omega^0(\mathcal{N}))$ given by $\tau_\rho(X) = \rho(X_\mathcal{N}) + \mu(X)$ determines a linear map $\tau\colon H^2_{\mathcal{G}}(\mathcal{N}) \to H^1(\mathrm{Lie}\,\mathcal{G}, \Omega^0(\mathcal{N}))$ which is independent of the form ρ chosen, and that we call the transgression map τ. If $\mathcal{G}$ is connected, then τ is injective.*

Proof The first part of the Proposition easily follows using that $L_{Y_\mathcal{N}}\mu(X) = \mu([Y, X])$ by the invariance of μ. We restrict ourselves to prove that τ is injective. By definition $[\tau_\rho] = 0$ on $H^1(\mathrm{Lie}\,\mathcal{G}, \Omega^0(\mathcal{N}))$ if and only if there exists $\beta \in \Omega^0(\mathcal{N})$ such that for every $X \in \mathrm{Lie}\,\mathcal{G}$ we have $\tau_\rho(X) = L_{X_\mathcal{N}}\beta = \iota_{X_\mathcal{N}}d\beta$. If we set $\rho' = \rho - d\beta$ then for every $X \in \mathrm{Lie}\,\mathcal{G}$ we have $d\rho' = \omega_0$, $\iota_{X_\mathcal{N}}\rho' = -\mu(X)$, $L_{X_\mathcal{N}}\rho' = 0$, *i.e.*, $\rho' \in \Omega^1(\mathcal{N})^{\mathcal{G}}$ and $d_c\rho' = \omega$.

Now we assume that $\pi\colon \mathcal{N} \to \mathcal{N}/\mathcal{G}$ is a principal $\mathcal{G}$-bundle. Then we can consider the more familiar transgression map defined as follows

Proposition 3 *Let $\underline{\omega} \in \Omega^2(\mathcal{N}/\mathcal{G})$ be a closed 2-form. If $\eta \in \Omega^1(\mathcal{N})$ is a form such that $\pi^*\underline{\omega} = d\eta$, then the map $t_\eta\colon \mathrm{Lie}\,\mathcal{G} \to \Omega^0(\mathcal{N})$, $t_\eta(X) = \eta(X_\mathcal{N})$ determines a linear map $t\colon H^2(\mathcal{N}/\mathcal{G}) \to H^1(\mathrm{Lie}\,\mathcal{G}, \Omega^0(\mathcal{N}))$, which is independent of the form η chosen, and that we call the transgression map t. If $\mathcal{G}$ is connected, then t is injective.*

Proof Again we only prove that t is injective. If $t_\eta = \delta\upsilon$ for certain $\upsilon \in \Omega^0(\mathcal{N})$, then $\eta(X_\mathcal{N}) = L_{X_\mathcal{N}}\upsilon$. We define $\eta' = \eta - d\upsilon$, and we have $d\eta' = \pi^*\underline{\omega}$, $\iota_{X_\mathcal{N}}\eta' = 0$, $L_{X_\mathcal{N}}\eta' = 0$. Hence η' is projectable onto a form $\underline{\eta'} \in \Omega^1(\mathcal{N}/\mathcal{G})$ and $d\underline{\eta'} = \underline{\omega}$.

Proposition 4 *If $\omega \in H^2_{\mathcal{G}}(\mathcal{N})$ and $\underline{\omega} = \mathrm{ChW}(\omega)$ then we have $\tau(\omega) = t(\underline{\omega})$.*

Proof If $\omega = \omega_0 + \mu$, by Proposition 1 we have $\omega = \pi^*\underline{\omega} + d_c\alpha$ for some $\alpha \in \Omega^1_{\mathcal{G}}(\mathcal{N}) = \Omega^1(\mathcal{N})^{\mathcal{G}}$, i.e. $\omega_0 = \pi^*\underline{\omega} + d\alpha$ and $\mu(X) = -\alpha(X_\mathcal{N})$.

Let $\eta \in \Omega^1(\mathcal{N})$ be a form such that $\pi^*\underline{\omega} = d\eta$. If we set $\rho = \eta + \alpha$ then $\omega_0 = d\rho$ and for every $X \in \mathrm{Lie}\,\mathcal{G}$ we have $\tau_\rho(X) = t_\eta(X)$.

3 Local Equivariant Cohomology

Let $p\colon E \to M$ be a bundle over a compact, oriented n-manifold M without boundary. We denote by $J^r E$ its r-jet bundle, by $J^\infty E$ the infinite jet bundle and by $\Gamma(E)$ be the manifold of global sections of E (assumed to be not empty).

We denote by $\mathrm{Proj}\,E$ the space of projectable diffeomorphism of E, and by $\mathrm{Proj}^+ E$ the subgroup of elements preserving the orientation of M. The space of projectable vector fields on E is denoted by $\mathrm{proj}\,E$. We consider the natural actions of $\mathrm{Proj}\,E$ on $J^\infty E$ and $\Gamma(E)$.

Let $\mathrm{j}^\infty\colon M \times \Gamma(E) \to J^\infty E, \mathrm{j}^\infty(x,s) = j^\infty_x s$ be the evaluation map. We define a map $\Im\colon \Omega^{n+k}(J^\infty E) \to \Omega^k(\Gamma(E))$, by $\Im[\alpha] = \int_M (\mathrm{j}^\infty)^* \alpha$, for $\alpha \in \Omega^{n+k}(J^\infty E)$. The map $\Im$ commutes with the exterior differential and is $\mathrm{Proj}^+ E$-equivariant (see [9]). We define the space of local k-forms on $\Gamma(E)$, as the image of the map $\Im$, i.e. $\Omega^k_{\mathrm{loc}}(\Gamma(E)) = \Im(\Omega^{n+k}(J^\infty E)) \subset \Omega^k(\Gamma(E))$. The cohomology $H^\bullet_{\mathrm{loc}}(\Gamma(E))$ of the complex $(\Omega^\bullet_{\mathrm{loc}}(\Gamma(E)), d)$ is called the local cohomology of $\Gamma(E)$. We have $H^k_{\mathrm{loc}}(\Gamma(E)) \cong H^{n+k}(E)$ for $k > 0$ (see [10]).

Let $\mathcal{G}$ be a Lie group acting on E by elements $\mathrm{Proj}^+ E$. In order to define an adequate notion of local equivariant cohomology we made the following

Assumption 1 We assume that $\mathrm{Lie}\,\mathcal{G}$ is isomorphic to the space of sections of a Lie algebroid $V \to M$, i.e. $\mathrm{Lie}\,\mathcal{G} \cong \Gamma(V)$. We also assume that the map $\mathrm{Lie}\,\mathcal{G} \cong \Gamma(V) \to \mathrm{proj}\,E$, $X \mapsto X_E$ is a differential operator. Finally, in the definition of $\mathcal{G}$-equivariant cohomology $H^{n+k}_{\mathcal{G}}(J^\infty E)$, we assume that the polynomial maps $\alpha\colon \mathrm{Lie}\,\mathcal{G} \to \Omega^\bullet(J^\infty E)$ are differential operators.

We extend the integration operator to a map $\Im\colon \Omega^{n+k}_{\mathcal{G}}(J^\infty E) \to \Omega^k_{\mathcal{G}}(\Gamma(E))$, by setting $(\Im[\alpha])(X) = \Im[\alpha(X)]$ for every $\alpha \in \Omega^{n+k}_{\mathcal{G}}(J^\infty E)$, $X \in \mathrm{Lie}\,\mathcal{G}$. The map $\Im$ commutes with the Cartan differential and induces a homomorphism in equivariant cohomology $\Im\colon H^{n+k}_{\mathcal{G}}(J^\infty E) \to H^k_{\mathcal{G}}(\Gamma(E))$ (see [9]).

We define the space of local $\mathcal{G}$-equivariant k-forms by

$$\Omega^k_{\mathcal{G},\mathrm{loc}}(\Gamma(E)) = \Im(\Omega^{n+k}_{\mathcal{G}}(J^\infty E)) \subset \Omega^k_{\mathcal{G}}(\Gamma(E)). \tag{2}$$

The local $\mathcal{G}$-equivariant cohomology of $\Gamma(E)$, $H^\bullet_{\mathcal{G},\mathrm{loc}}(\Gamma(E))$, is defined as the cohomology of the complex $(\Omega^\bullet_{\mathcal{G},\mathrm{loc}}(\Gamma(E)), d_c)$.

4 Application to Local Anomaly Cancellation

Let us define the BRST cohomology (see [6, 16]). Recall (see Assumption 1) that we assume $\mathrm{Lie}\,\mathcal{G} \cong \Gamma(V)$ for some vector bundle $V \to M$. A map $\alpha\colon \bigwedge^k \mathrm{Lie}\,\mathcal{G} \to \Omega^0_{\mathrm{loc}}(\Gamma(E))$ is said to be local if there exists a differential operator $A\colon \bigwedge^k \mathrm{Lie}\,\mathcal{G} \to \Omega^n(J^\infty E)$ such that $\alpha(X_1, \dots, X_k) = \Im[A(X_1, \dots, X_k)]$ for every $X_1, \dots, X_k \in \mathrm{Lie}\,\mathcal{G}$. We denote by $\Omega^k_{\mathrm{loc}}(\mathrm{Lie}\,\mathcal{G}, \Omega^0_{\mathrm{loc}}(\Gamma(E))$ the space of local k-forms on $\mathrm{Lie}\,\mathcal{G}$ with values on $\Omega^0_{\mathrm{loc}}(\Gamma(E))$. The differential δ on the complex $\Omega^\bullet(\mathrm{Lie}\,\mathcal{G}, \Omega^0_{\mathrm{loc}}(\Gamma(E))$ induces a differential on $\Omega^\bullet_{\mathrm{loc}}(\mathrm{Lie}\,\mathcal{G}, \Omega^0_{\mathrm{loc}}(\Gamma(E))$. The corresponding cohomology $H^\bullet_{\mathrm{loc}}(\mathrm{Lie}\,\mathcal{G}, \Omega^0_{\mathrm{loc}}(\Gamma(E)))$ is called the BRST cohomology. We assume that $H^2(\Gamma(E)) = H^1(\Gamma(E)) = 0$ and also that $H^2_{\mathrm{loc}}(\Gamma(E)) = H^1_{\mathrm{loc}}(\Gamma(E)) = 0$.

Proposition 5 *The restriction of the transgression map τ to $H^2_{\mathcal{G},\mathrm{loc}}(\Gamma(E))$ takes values on the BRST cohomology $H^1_{\mathrm{loc}}(\mathrm{Lie}\,\mathcal{G}, \Omega^0_{\mathrm{loc}}(\Gamma(E)))$. The map $\tau\colon H^2_{\mathcal{G},\mathrm{loc}}(\Gamma(E)) \to H^1_{\mathrm{loc}}(\mathrm{Lie}\,\mathcal{G}, \Omega^0_{\mathrm{loc}}(\Gamma(E)))$ is injective for $\mathcal{G}$ connected.*

Proof Let $\omega = \omega_0 + \mu \in \Omega^2_{\mathcal{G},\mathrm{loc}}(\Gamma(E))$ be a closed local $\mathcal{G}$-equivariant 2-form. As $H^2_{\mathrm{loc}}(\Gamma(E)) = 0$, we have $\omega_0 = d\rho$, for certain $\rho \in \Omega^1_{\mathrm{loc}}(\Gamma(E))$. By our assumption in the definition of local equivariant cohomology (Assumption 1), the map $\tau_\rho\colon \mathrm{Lie}\,\mathcal{G} \to \Omega^0_{\mathrm{loc}}(\Gamma(E))$, $\tau_\rho(X) = \rho(X_{\Gamma(E)}) + \mu(X) \in \Omega^0_{\mathrm{loc}}(\Gamma(E))$ is a local map. The injectiveness of τ follows from Proposition 2.

Assume that $\Gamma(E) \to \Gamma(E)/\mathcal{G}$ is a principal $\mathcal{G}$-bundle. Then we define the local cohomology by $H^k_{\mathrm{loc}}(\Gamma(E)/\mathcal{G}) = \mathrm{ChW}(H^k_{\mathcal{G},\mathrm{loc}}(\Gamma(E)))$. By Proposition 4 we have the following

Proposition 6 *Let $\omega \in \Omega^2_{\mathcal{G},\mathrm{loc}}(\Gamma(E))$ be a closed local $\mathcal{G}$-equivariant 2-form and let $\underline{\omega} = \mathrm{ChW}(\omega)$. Then we have $\tau(\omega) = t(\underline{\omega})$, and in particular we conclude that $t(\underline{\omega}) \in H^1_{\mathrm{loc}}(\mathrm{Lie}\,\mathcal{G}, \Omega^0_{\mathrm{loc}}(\Gamma(E)))$. Moreover, the following conditions are equivalent*

(a) $[\omega] = 0$ *on* $H^2_{\mathcal{G},\mathrm{loc}}(\Gamma(E))$.
(b) $[\underline{\omega}] = 0$ *on* $H^2_{\mathrm{loc}}(\Gamma(E)/\mathcal{G})$.
(c) $[\tau(\omega)] = [t(\underline{\omega})] = 0$ *on* $H^1_{\mathrm{loc}}(\mathrm{Lie}\,\mathcal{G}, \Omega^0_{\mathrm{loc}}(\Gamma(E)))$.

Proposition 6 applied to $\underline{\omega} = c_1(\det \mathrm{Ind} D)$ shows that our definition of $H^2_{\mathrm{loc}}(\Gamma(E)/\mathcal{G})$ solves Singer's problem. We also note that if $\omega \in \Omega^2_{\mathcal{G},\mathrm{loc}}(\Gamma(E))$ is closed, the form $\underline{\omega} \in \Omega^2(\Gamma(E)/\mathcal{G})$ determining the class $\mathrm{ChW}([\omega])$ could contain non-local terms, as $\underline{\omega}$ depends on the curvature of a connection Θ on the principal $\mathcal{G}$-bundle $\Gamma(E) \to \Gamma(E)/\mathcal{G}$, and Θ usually contains non-local terms. This fact explains the appearance of non-local terms on the expression of the curvature of $\det \mathrm{Ind} D$ commented on the Introduction.

References

1. Álvarez, O., Singer, I., Zumino, B.: Gravitational anomalies and the family's index theorem. Commun. Math. Phys. **96**, 409–417 (1984)
2. Atiyah, M.F., Singer, I.: Dirac operators coupled to vector potentials. Proc. Natl. Acad. Sci. USA **81**, 2597–2600 (1984)
3. Berline, N., Getzler, E., Vergne, M.: Heat Kernels and Dirac Operators, Springer, Berlin (1992)
4. Bertlmann, R.A.: Anomalies in Quantum Field Theory. Oxford University Press, Oxford (2000)
5. Blau, M.: Wess-Zumino terms and the geometry of the determinant line bundle. Phys. Lett. **209** B, 503–506 (1988)

6. Bonora, L., Cotta-Ramusino, P.: Some remarks on BRS transformations, anomalies and the cohomology of the Lie algebra of the group of gauge transformations. Commun. Math. Phys. **87**, 589–603 (1983)
7. Bonora, L., Cotta-Ramusino, P.: Consistent and covariant anomalies and local cohomology. Phys. Rev. D **33**(3), 3055–3059 (1986)
8. Dubois-Violette, M., Henneaux, M., Talon, M., Viallet, C.: General solution of the consistency equation. Phys. Lett. B **289**, 361–367 (1992)
9. Pérez, R.F.: Equivariant characteristic forms in the bundle of connections. J. Geom. Phys. **54**, 197–212 (2005)
10. Pérez, R.F.: Local cohomology and the variational bicomplex. Int. J. Geom. Methods Mod. Phys. **5** (2008), 587–604
11. Pérez, R.F.: Local anomalies and local equivariant cohomology. Comm. Math. Phys. **286**, 445–458 (2009)
12. Pérez, R.F., Masqué, J,M.: Natural connections on the bundle of Riemannian metrics. Monatsh. Math. **155**, 67–78 (2008)
13. Guillemin, V., Sternberg, S.: Supersymmetry and Equivariant de Rham Theory. Springer, Berlin (1999)
14. Mañes, J., Stora, R., Zumino, B.: Algebraic study of chiral anomalies. Comm. Math. Phys. **102**, 157–174 (1985)
15. Martellini, M., Reina, C.: Some remarks on the index theorem approach to anomalies. Ann. Inst. H. Poincarè **113**, 443–458 (1985)
16. Schmid, R.: Local cohomology in gauge theories, BRST transformations and anomalies. Differential Geom. Appl. **4**(2), 107–116 (1994)
17. Singer, I.M.: Families of Dirac operators with applications to physics, The mathematical heritage of Élie Cartan (Lyon, 1984). Astérisque (1985), Numero Hors Serie, 323–340

Classes of Nonlinear Filters for Stream Ciphers

Amparo Fúster-Sabater and Fausto Montoya Vitini

Queremos dedicar este trabajo al profesor Jaime Muñoz Masqué con motivo de su 65° cumpleaños. Gracias por todo Jaime.

Abstract Long period, good statistical properties and large linear complexity are necessary conditions that every cryptographic sequence must satisfy. In this work, an algebraic method to compute classes of nonlinear filters with large linear complexity has been proposed. Two filter operations (addition and shifting operations) are performed to give rise to a complete class of nonlinear filters adequate for cryptographic purposes. The procedure here developed is simple, efficient and can be carried out at the price of minimal computational operations. Different filter representations have been systematically addressed.

Keywords Linear complexity · Sequence generator · Filter function · Cryptography

1 Introduction

A stream cipher cryptosystem consists of a short key and a public algorithm or sequence generator. The output sequence generated by such a generator (*keystream sequence*) is XORed with the plaintext (in emission) to obtain the ciphertext or with

This article is dedicated to Professor Jaime Muñoz Masqué on his 65th birthday. Jaime, we thank you for everything.

A. Fúster-Sabater (✉) · F.M. Vitini
Instituto de Tecnologías Físicas y de la Información, CSIC, 28006 Madrid, Spain
e-mail: amparo@iec.csic.es

F.M. Vitini
e-mail: fausto@iec.csic.es

M. Castrillón López et al. (eds.), *Geometry, Algebra and Applications: From Mechanics to Cryptography*, Springer Proceedings in Mathematics & Statistics 161, DOI 10.1007/978-3-319-32085-4_10

the ciphertext (in reception) to recover the original plaintext. References [9–11, 18] provide a solid introduction to the study of stream ciphers.

At the present moment, stream ciphers are the fastest among the encryption procedures so they are implemented in many engineering applications e.g. the encryption algorithm RC4 [12] used in Wired Equivalent Privacy (WEP) as a part of the 802.11 standard, the recent proposals HC-128 or Rabbit coming from the eSTREAM Project [16] that are included in the latest release versions of CyaSSL (lightweight open source embedded implementation of the SSL/TLS protocol [19]) or the J3Gen, a promising pseudo random number generator for low-cost passive Radio Frequency Identification (RFID) tags [14].

Typically, keystream generators are based on maximal-length Linear Feedback Shift Registers (LFSRs) [3] whose output sequences, the so-called m-sequences, are combined in a nonlinear way (e.g. by means of nonlinearly filtering or combination, irregular decimation, modelling with cellular automata, introduction of typical elements from block ciphers, etc) to produce sequences of cryptographic application. One general technique for building keystream generators is to use a nonlinear filter, i.e. a nonlinear function applied to the stages of a single maximal-length LFSR [2, 11]. That is the output sequence is generated as the image of a nonlinear Boolean function F in the LFSR stages.

Desirable properties for such sequences can be enumerated as follows: (a) Long Period, (b) Good statistical properties, (c) Large Linear Complexity (LC). Period and statistical properties of the filtered sequences are characteristics deeply studied in the literature, see the references above mentioned as well as [1, 13, 15]. In addition, such sequences have to pass a battery of tests (DIEHARD tests [6], Tuftests [7]) to be accepted as cryptographic sequences. Regarding the third requirement, linear complexity of a sequence is defined as the length of the shortest LFSR able to generate such a sequence. In cryptographic terms, LC must be as large as possible in order to prevent the application of the Berlekamp-Massey algorithm [8]. A recommended value for LC is about half the sequence period. Although the exact value of the linear complexity attained by any filtered sequence is still an open problem [5], in this work a method of computing the whole class of filtering functions with a linear complexity adequate for cryptographic purposes is proposed. The method is based on the handling of nonlinear filters by means of algebraic operations.

2 Main Concepts and Basic Notation

Specific notation and different basic concepts to develop such a computing method are introduced.

Maximal-sequence (m-sequence). Let $\{s_n\}$ be the binary output sequence of a maximal-length LFSR of L stages, that is a LFSR whose characteristic polynomial $P(x) = \sum_{j=0}^{L} p_j x^j$ with $p_j \in \{0, 1\}$ is primitive of degree L, see [18]. In that case, the output sequence is a m-sequence of period $2^L - 1$. The sequence $\{s_n\}$ satisfies the linear recursion:

$$\sum_{j=0}^{L} p_j \, s_{n+j} = 0.$$

The roots of $P(x)$ are of the form α^{2^i} $(i = 0, 1, \ldots, L-1)$ where α is a primitive element in $GF(2^L)$ that is an extension of the binary field $GF(2)$ with 2^L elements [4]. The generic term of the sequence $\{s_n\}$ can be written by means of the roots of $P(x)$ as [3]:

$$s_n = Tr(C\,\alpha^n) = \sum_{j=0}^{L-1} (C\alpha^n)^{2^j}, \quad n \geq 0 \tag{1}$$

where $C \in GF(2^L)$. If $C = 1$, then $\{s_n\}$ it is said to be in its *characteristic phase.*

Nonlinear filter It is a Boolean function $F(x_0, x_1, \ldots, x_{L-1})$ in L variables of degree k. For a subset $A = \{a_0, a_1, \ldots, a_{r-1}\}$ of $\{0, 1, \ldots, L-1\}$ with $r \leq k$, the notation $x_A = x_{a_0} x_{a_1} \ldots x_{a_{r-1}}$ is used. The Boolean function can be written as [17]:

$$F(x_0, x_1, \ldots, x_{L-1}) = \sum_{A} c_A \, x_A, \tag{2}$$

where $c_A \in \{0, 1\}$ are binary coefficients and the summation is taken over all subsets A of $\{0, 1, \ldots, L-1\}$.

Filtered sequence. The sequence $\{z_n\}$ is the keystream or output sequence of the nonlinear filter F applied to the L stages of the LFSR. The keystream bit z_n is computed by selecting bits from the m-sequence such that

$$z_n = F(s_n, s_{n+1}, \ldots, s_{n+L-1}).$$

Cyclotomic coset. Let Z_{2^L-1} denote the set of integers $[1, \ldots, 2^L - 1]$. An equivalence relation R is defined on its elements $q_1, q_2 \in Z_{2^L-1}$ such as follows: $q_1 R\, q_2$ if there exists an integer j, $0 \leq j \leq L-1$, such that

$$2^j \cdot q_1 = q_2 \bmod 2^L - 1.$$

The resultant equivalence classes into which Z_{2^L-1} is partitioned are called the *cyclotomic cosets* mod $2^L - 1$, see [3]. All the elements q_i of a cyclotomic coset have the same number of 1's in their binary representation; this number is called the *coset weight.* The leader element, E, of every coset is the smallest integer in such an equivalence class. Moreover, the cardinal of any coset is L or a proper divisor of L.

Characteristic polynomial of a cyclotomic coset. It is a polynomial $P_E(x)$ defined by $P_E(x) = (x + \alpha^E)(x + \alpha^{2E}) \ldots (x + \alpha^{2^{(r-1)}E})$, where the degree r $(r \leq L)$ of $P_E(x)$ equals the cardinal of the cyclotomic coset E.

Characteristic sequence of a cyclotomic coset. It is a binary sequence $\{S_n^E\}$ defined by the expression $\{S_n^E\} = \{\alpha^{En} + \alpha^{2En} + \cdots + \alpha^{2^{(r-1)}En}\}$ with $n \geq 0$. Recall that the sequence $\{S_n^E\}$ is in its characteristic phase and satisfies the linear recurrence relationship given by $P_E(x)$, see [4].

3 Nonlinear Filter Representations and Equivalence Classes

The Eq. (2) describes the Algebraic Normal Form (ANF) of a nonlinear filter $F(s_n, s_{n+1}, \ldots, s_{n+L-1})$. This representation of Boolean functions is currently used by the designer of nonlinear filters as he can handle the degree and particular form of the function. Nevertheless, the ANF of nonlinear filters do not give information on the linear complexity of the filtered sequence. In this sense, a different representation closely related to the linear complexity is required. Thus, a new nonlinear filter representation is introduced. Indeed, if all the variables s_{n+j} $(0 \leq j \leq L-1)$ of F are substituted by their corresponding expressions in (1) and the resulting terms grouped, then the term z_n of the filtered sequence $\{z_n\}$ can be written as:

$$\begin{array}{c} z_n = F(s_n, s_{n+1}, \ldots, s_{n+L-1}) = \\ C_{E_1}\alpha^{E_1 n} + (C_{E_1}\alpha^{E_1 n})^2 + \cdots + (C_{E_1}\alpha^{E_1 n})^{2^{(r_1-1)}} + \\ \vdots \\ C_{E_i}\alpha^{E_i n} + (C_{E_i}\alpha^{E_i n})^2 + \cdots + (C_{E_i}\alpha^{E_i n})^{2^{(r_i-1)}} + \\ \vdots \\ C_{E_N}\alpha^{E_N n} + (C_{E_N}\alpha^{E_N n})^2 + \cdots + (C_{E_N}\alpha^{E_N n})^{2^{(r_N-1)}}, \end{array} \quad (3)$$

where r_i is the cardinal of coset E_i, the subindex i ranges in the interval $1 \leq i \leq N$ and N is the number of cosets of weight $\leq k$. Thus, a nonlinear filter $F(s_n, s_{n+1}, \ldots, s_{n+L-1})$ can be represented in terms of the N characteristic sequences $\{S_n^{E_i}\}$ that appear in this sequential decomposition shown in Eq. (3).

Note that the ith row of (3) corresponds to the nth-term of the sequence $\{C_{E_i}\alpha^{E_i n} + (C_{E_i}\alpha^{E_i n})^2 + \cdots + (C_{E_i}\alpha^{E_i n})^{2^{(r_i-1)}}\}$, that is the characteristic sequence $\{S_n^{E_i}\}$ where the coefficient $C_{E_i} \in GF(2^{r_i})$ determines the starting point with reference to its characteristic phase. If $C_{E_i} = 0$, then the corresponding cyclotomic coset E_i would be *degenerate*. Otherwise, the coset is *nondegenerate*. Linear complexity of the filtered sequence is related with the number of coefficients $C_{E_i} \neq 0$ as the contribution of any nondegenerate coset to LC equals the cardinal of such a coset [18]. In this way, the above sequential representation already provides quantitative information on the LC of the filtered sequence.

Now, we can introduce a third nonlinear filter representation that is a simplification of the previous one. In fact, a nonlinear filter $F(s_n, s_{n+1}, \ldots, s_{n+L-1})$ can be also represented by a N-tuple of coefficients $(C_{E_1}, C_{E_2}, \ldots, C_{E_N})$ where $C_{E_i} \in GF(2^{r_i})$ and N is as before the number of cosets of weight $\leq k$. This representation is particularly useful as allows us to treat separately the distinct cosets.

In brief, a nonlinear filter can be represented in three different ways:

1. The traditional Algebraic Normal Form.
2. The sum of its characteristic sequences.
3. A N-tuple of coefficients that define the starting point of such characteristics sequences.

The idea of grouping nonlinear filters in equivalence classes for their analysis and handling has been already treated in the literature, see [17]. In this section, an equivalence relationship specific to design filters with guaranteed LC is proposed.

Let G be the set of the kth-order nonlinear filters applied to a single LFSR of length L. We can group the elements of G producing the filtered sequence $\{z_n\}$ or a shifted version of $\{z_n\}$, denoted by $\{z_n\}^*$. From equation (3), it is clear that if we substitute C_{E_i} for $C_{E_i} \cdot \alpha^{E_i}$ $(1 \leq i \leq N)$, then we will obtain $\{z_{n+1}\}$. In general,

$$C_{E_i} \rightarrow C_{E_i} \cdot \alpha^{jE_i} \;\; \forall i \;\; \Rightarrow \{z_n\} \rightarrow \{z_{n+j}\}.$$

This fact enables us to define an equivalence relationship $\sim$ on the set G as follows: $F \sim F'$ with $F, F' \in G$ if they generate shifted versions of the same filtered sequence $\{z_n\}$,

$$\{F(s_n, \ldots, s_{n+L-1})\} = \{z_n\} \text{ and } \{F'(s_n, \ldots, s_{n+L-1})\} = \{z_n\}^*.$$

It is easy to see that the relation defined above is an equivalence relationship and that the number of filters in each equivalence class equals the period of the filtered sequence. Making use of the third representation for nonlinear filters (N-tuple of coefficients), we see that the coefficients associated with F, F', notated (C_{E_i}) and (C'_{E_i}) respectively, satisfy for an integer j

$$C'_{E_i} = C_{E_i} \cdot \alpha^{j\,E_i} \quad (1 \leq i \leq N). \tag{4}$$

4 Computing Nonlinear Filters with a Guaranteed *LC*

Previously to the method's description, several results that will be used in the computation are introduced.

Definition 1 Two nonlinear filters F_0 and F_1 in the same equivalence class are consecutive if they satisfy the Eq. (4) with $j = 1$ or equivalently

$$F_1(s_n, \ldots, s_{n+L-1}) = F_0(s_{n+1}, \ldots, s_{n+L}).$$

Moving from filter F_0 to filter F_1 in the three forms of representation means:

1. An increment by 1 in all the sub-indices of the ANF representation followed (if necessary) by a substitution of s_{n+L} by $\sum_{j=0}^{L-1} p_j\, s_{n+j}$.

2. A simultaneous cyclic shift of all the characteristic sequences $\{S_n^{E_i}\}$ in the characteristic sequence-based representation.
3. The product of each coefficient by its corresponding factor α^{E_i} $(1 \le i \le N)$ in the N-tuple representation.

Let $E_1, E_2, \ldots, E_M$ be the leaders of the M nondegenerate cosets of weight $\le k$ in $\{z_n\}$ and $r_1, r_2, \ldots, r_M$ their corresponding cardinals. Two different Lemmas that allow one to handle the different filters of an equivalence class can be pointed out.

Lemma 1 *If p nonlinear filters in the same equivalence class are chosen*

$$(C_{E_i}), (C_{E_i} \cdot \alpha^{q_1 E_i}), (C_{E_i} \cdot \alpha^{q_2 E_i}), \ldots, (C_{E_i} \cdot \alpha^{q_{p-1} E_i}), \tag{5}$$

$(q_1, q_2, \ldots, q_{p-1})$ *being integers in such a way that no characteristic polynomial* $P_{E_i}(x)$ $(1 \le i \le M)$ *divides the polynomial*

$$Q(x) = (1 + x^{q_1} + \cdots + x^{q_{p-1}}),$$

then the nonlinear filter characterized by the coefficients

$$\tilde{C}_{E_i} = C_{E_i}(1 + \alpha^{q_1 E_i} + \cdots + \alpha^{q_{p-1} E_i}) \ \ (1 \le i \le M)$$

preserves the same cosets E_i *as those of the filters defined in (5).*

Proof The result follows from the fact that the coefficients of the new nonlinear filter verify

$$\tilde{C}_{E_i} = C_{E_i}(1 + \alpha^{q_1 E_i} + \cdots + \alpha^{q_{p-1} E_i}) \neq 0 \ \ (1 \le i \le M)$$

as no α^{E_i} is a root of $Q(x)$. □

Therefore an easy way to guarantee the presence of all the cosets E_i in the new filter is just summing $p \le r_{min}$ consecutive nonlinear filters in the same equivalence class (r_{min} being the least cardinal of all the cosets E_i) as $deg\, Q(x) < deg\, P_{E_i}(x)$ $(1 \le i \le M)$.

Lemma 2 *The sum of nonlinear filters satisfying the conditions of Lemma 1 gives rise to a new nonlinear filter in a different equivalence class.*

Proof We proceed by contradiction. Suppose that the new filter belongs to the same equivalence class. Then,

$$\tilde{C}_{E_i} = C_{E_i}(1 + \alpha^{q_1 E_i} + \cdots + \alpha^{q_{p-1} E_i}) = C_{E_i} \cdot \alpha^{j E_i} \ \ (1 \le i \le M). \tag{6}$$

Assume without loss of generality that coset E_1 = coset 1. Therefore, according to Eq. (6)

$$(1 + \alpha^{q_1} + \cdots + \alpha^{q_{p-1}}) = \alpha^j$$

and consequently

$$(1+\alpha^{q_1 E_i}+\cdots+\alpha^{q_{p-1}E_i})=\alpha^{jE_i}\ \ (2\le i\le M).$$

Thus, it follows that

$$(1+\alpha^{q_1}+\cdots+\alpha^{q_{p-1}})^{E_i}=(1+\alpha^{q_1 E_i}+\cdots+\alpha^{q_{p-1}E_i})\ \ (2\le i\le M).$$

Nevertheless, it is a well known fact that in $GF(2^L)$ this equality only holds for E_i of the form 2^m (i.e. the elements of coset 1) but not for the leaders of any coset $E_i \neq$ coset 1. □

Both lemmas allow us to construct a new equivalence class that preserves the same cosets as those of the initial class.

From the previous results, an easy method of computing all the nonlinear filters that guarantee the cosets of weight k is given. Indeed, we start from a filter with a unique term product of k equidistant phases of the form:

$$F_0(s_n, s_{n+1}, \ldots, s_{n+L-1}) = s_n s_{n+\delta} s_{n+2\delta} \ldots s_{n+(k-1)\delta} \tag{7}$$

with $1 \le k \le L$ and $gcd(\delta, 2^L-1)=1$. In the sequel, we will focus exclusively on the N_k cosets of weight k making use of the N_k-tuple representation.

Given F_0 in (7), the computation of its N_k-tuple is carried out as follows. Let $E=2^{e_0}+2^{e_1}+\cdots+2^{e_{k-1}}$ be the leader element of an arbitrary coset of weight k where e_i $(0\le i\le k-1)$ are integers. According to Eqs. (2), (3) and grouping terms, the coefficient C_E for F_0 in the N_k-tuple representation, denoted by C_E^0, is given by the determinant

$$C_E^0 = \begin{vmatrix} \alpha^{0\cdot 2^{e_0}} & \alpha^{\delta\, 2^{e_0}} & \alpha^{2\,\delta\, 2^{e_0}} & \ldots & \alpha^{(k-1)\,\delta\, 2^{e_0}} \\ \alpha^{0\cdot 2^{e_1}} & \alpha^{\delta\, 2^{e_1}} & \alpha^{2\,\delta\, 2^{e_1}} & \ldots & \alpha^{(k-1)\,\delta\, 2^{e_1}} \\ & \ldots & & \ldots & \ldots \\ \alpha^{0\cdot 2^{e_{k-1}}} & \alpha^{\delta\, 2^{e_{k-1}}} & \alpha^{2\,\delta\, 2^{e_{k-1}}} & \ldots & \alpha^{(k-1)\,\delta\, 2^{e_{k-1}}} \end{vmatrix},$$

or equivalently,

$$C_E^0 = \begin{vmatrix} 1 & \lambda_0 & \lambda_0^2 & \ldots & \lambda_0^{(k-1)} \\ 1 & \lambda_1 & \lambda_1^2 & \ldots & \lambda_1^{(k-1)} \\ & \ldots & & \ldots & \ldots \\ 1 & \lambda_{k-1} & \lambda_{k-1}^2 & \ldots & \lambda_{k-1}^{(k-1)} \end{vmatrix} = \Pi(\lambda_i+\lambda_j),$$

with $\lambda_i=\alpha^{\delta\,2^{e_i}}$, $\lambda_j=\alpha^{\delta\,2^{e_j}}$ $(0\le i<j\le k-1)$. Thus, each coefficient C_E^0 is a Vandermonde determinant that can be easily computed as well as it is guaranteed to be different from 0. Thus, the cosets of weight k for the filter F_0 are nondegenerate and their contribution to the linear complexity equals $\binom{L}{k}$, see [18]. Next, the coefficient of coset E for F_1 in the N_k-tuple representation, denoted by C_E^1, is given by the

determinant

$$C_E^1 = \begin{vmatrix} \alpha^{1\cdot 2^{e_0}} & \alpha^{(\delta+1)\,2^{e_0}} & \alpha^{(2\,\delta+1)\,2^{e_0}} & \dots & \alpha^{((k-1)\,\delta+1)\,2^{e_0}} \\ \alpha^{1\cdot 2^{e_1}} & \alpha^{(\delta+1)\,2^{e_1}} & \alpha^{(2\,\delta+1)\,2^{e_1}} & \dots & \alpha^{((k-1)\,\delta+1)\,2^{e_1}} \\ & \dots & & \dots & \dots \\ \alpha^{1\cdot 2^{e_{k-1}}} & \alpha^{(\delta+1)\,2^{e_{k-1}}} & \alpha^{(2\,\delta+1)\,2^{e_{k-1}}} & \dots & \alpha^{((k-1)\,\delta+1)\,2^{e_{k-1}}} \end{vmatrix}$$

thus,

$$C_E^1 = \alpha^{2^{e_0}} \cdot \alpha^{2^{e_1}} \dots \alpha^{2^{e_{k-1}}} \cdot C_E^0 = \alpha^E \cdot C_E^0.$$

At the same time, the coefficient of coset E for the filter $F_0 + F_1$, denoted by C_E^{01}, is

$$C_E^{01} = C_E^0 + C_E^1 = C_E^0(1 + \alpha^E).$$

The key idea in this construction method is shifting the filter $F_0 + F_1$ through its equivalence class and summing it with F_0 in order to cancel the successive components of its N_k-tuple. The procedure is sketched in algorithm 1. After a succession of sums and shiftings, the final result is:

1. A set of N_k basic nonlinear filters of the form $(0, 0, \dots, d_i, \dots, 0, 0)$ $(1 \leq i \leq N_k)$ with $d_i \in GF(2^L)$, $d_i \neq 0$ where each filter includes a unique coset E_i of weight k.
2. Their corresponding ANF representations.

The combination of all these basic filters with d_i $(1 \leq i \leq N_k)$ ranging in $GF(2^L)$ (with the corresponding ANF representations) gives rise to all the possible terms of order k that preserve the cosets of weight k. Later, the addition of terms of order $< k$ in ANF permits the generation of all the nonlinear filters of order k that guarantee a linear complexity $LC \geq \binom{L}{k}$.

Algorithm 1. Computation of nonlinear filters with a unique k-weighted coset

Input:

The pair (L, k), N_k, LFSR characteristic polynomial $P(x)$, filter F_0 in ANF

Compute the N_k-tuple for filter F_0, $(C_{E_i}^0)$, $(1 \leq i \leq N_k)$;
for i = $N_k - 1$ **to** 2 **do**
  Compute the N_k-tuples for filters F_1 and $F_0 + F_1$, $(C_{E_i}^1)$ and $(C_{E_i}^{01})$, respectively;
  Shift $(C_{E_i}^{01})$ through its equivalence class until the $i - th$ component equals $C_{E_i}^0$;
  Sum $(C_{E_i}^{01}) + (C_{E_i}^0)$ getting a N_k-tuple with 0 at the $i - th$ component;
  $(C_{E_i}^0) \leftarrow (C_{E_i}^{01}) + (C_{E_i}^0)$
end for

Output:

N_k basic filters of the form $(0, 0, \dots, d_i, \dots, 0, 0)$ $(1 \leq i \leq N_k)$ with $d_i \in GF(2^L)$, $d_i \neq 0$
Their corresponding ANF representations.

4.1 Discussion of the Method

Regarding the previous method, distinct considerations must be taken into account.

Recall that the construction method above described to compute the basic filters $(0, 0, \ldots, d_i, \ldots, 0, 0)$, $d_i \neq 0$ involves very simple operations:

- Sum operation: that is reduced to a sum of filters in the ANF representation or to a sum of elements C_{E_i} in the extended field $GF(2^L)$ that expressed in binary representation is just an exclusive OR operation.
- Shifting operation through an equivalence class: that means for each shifting an increment by 1 in all the indexes in the ANF representation or the multiplication of powers of α by their corresponding factors α^{E_i} in the N_k-tuple representation.

Consequently, the efficiency of the computation method is quite evident. In brief, we provide one with the complete class of nonlinear filters with $LC \geq \binom{L}{k}$ at the price of minimal computational operations.

In the case that the presence of more cosets of weight $<k$ were guaranteed, the procedure here described continues being applicable just enlarging the coefficient vector to new components corresponding to those new guaranteed cosets in the N-tuple representation. Let us now see an illustrative example.

4.2 A Numerical Example

Let F_0 be a nonlinear filter of third order applied to the stages of a LFSR of length 5 and characteristic polynomial $P(x) = x^5 + x^3 + 1$, where α is a root of $P(x)$ so that $\alpha^5 = \alpha^3 + 1$. There are 2 cyclotomic cosets of weight 3: coset 7= $\{7, 14, 28, 25, 19\}$ and coset 11= $\{11, 22, 13, 26, 21\}$. The form of the filter with guaranteed cosets of weight 3 is $F_0(s_0, s_1, s_2) = s_0 s_1 s_2$. The algorithm previously described is applied.

INPUT: $(L, k) = (5, 3)$, $N_3 = 2$, LFSR characteristic polynomial $P(x) = x^5 + x^3 + 1$, filter $F_0 = s_0 s_1 s_2$ in ANF.

Computation of the 2-tuples for the filters $F_0, F_1, F_0 + F_1$:

- $F_0(s_0, s_1, s_2) = s_0 s_1 s_2 \rightarrow (C_7^0, C_{11}^0) = (\alpha^{20}, \alpha^{13})$.
- $F_1(s_0, s_1, s_2) = s_1 s_2 s_3 \rightarrow (C_7^1, C_{11}^1) = (\alpha^{20} \cdot \alpha^7, \alpha^{13} \cdot \alpha^{11}) = (\alpha^{27}, \alpha^{24})$.
- $F_0 + F_1 = s_0 s_1 s_2 + s_1 s_2 s_3 \rightarrow (C_7^{01}, C_{11}^{01}) = (\alpha^{20}, \alpha^{13}) + (\alpha^{27}, \alpha^{24}) = (\alpha^5, \alpha^5)$.

Computation of $(d_1, 0)$:

- Shifting of (C_7^{01}, C_{11}^{01}) through its equivalence class until $C_{11}^{01} = C_{11}^0 = \alpha^{13}$, that is (α^5, α^5) is shifted up to $(\alpha^{27}, \alpha^{13})$.
- Sum $(\alpha^{27}, \alpha^{13}) + (\alpha^{20}, \alpha^{13}) = (\alpha^5, 0)$, so that $(d_1, 0) = (\alpha^5, 0)$.

Computation of $(0, d_2)$:

Table 1 Class of nonlinear filters $(d_1, 0)$ with coset 7 exclusively

Filter	Algebraic normal form	Coeff.
F_0	$s_0s_1s_2 + s_0s_1s_3 + s_0s_1s_4 + s_0s_2s_3 + s_0s_2s_4 + s_0s_3s_4 + s_1s_2s_3 + s_1s_2s_4 + s_1s_3s_4$	$(\alpha^5, 0)$
F_1	$s_0s_1s_2 + s_0s_1s_3 + s_0s_1s_4 + s_0s_2s_3 + s_0s_2s_4 + s_1s_2s_4$	$(\alpha^{12}, 0)$
F_2	$s_0s_1s_2 + s_0s_1s_3 + s_0s_2s_3 + s_1s_2s_4 + s_1s_3s_4$	$(\alpha^{19}, 0)$
F_3	$s_0s_2s_3 + s_0s_2s_4 + s_1s_2s_3 + s_1s_2s_4 + s_1s_3s_4 + s_2s_3s_4$	$(\alpha^{26}, 0)$
F_4	$s_0s_1s_3 + s_0s_2s_3 + s_0s_2s_4 + s_0s_3s_4 + s_1s_3s_4$	$(\alpha^2, 0)$
F_5	$s_0s_1s_3 + s_0s_1s_4 + s_0s_2s_4 + s_1s_2s_4 + s_2s_3s_4$	$(\alpha^9, 0)$
F_6	$s_0s_1s_2 + s_0s_1s_3 + s_0s_2s_3 + s_0s_3s_4 + s_1s_2s_3 + s_1s_2s_4$	$(\alpha^{16}, 0)$
F_7	$s_0s_1s_4 + s_0s_2s_3 + s_1s_2s_3 + s_1s_2s_4 + s_2s_3s_4$	$(\alpha^{23}, 0)$
F_8	$s_0s_1s_2 + s_0s_2s_3 + s_0s_3s_4 + s_1s_2s_3 + s_1s_3s_4 + s_2s_3s_4$	$(\alpha^{30}, 0)$
F_9	$s_0s_1s_4 + s_0s_2s_4 + s_0s_3s_4 + s_1s_2s_3$	$(\alpha^6, 0)$
F_{10}	$s_0s_1s_2 + s_0s_1s_3 + s_0s_1s_4 + s_1s_2s_3 + s_1s_3s_4 + s_2s_3s_4$	$(\alpha^{13}, 0)$
F_{11}	$s_0s_1s_2 + s_0s_2s_4 + s_0s_3s_4 + s_1s_2s_4$	$(\alpha^{20}, 0)$
F_{12}	$s_0s_1s_3 + s_0s_1s_4 + s_0s_2s_3 + s_1s_2s_3 + s_1s_3s_4$	$(\alpha^{27}, 0)$
F_{13}	$s_0s_1s_2 + s_0s_2s_4 + s_1s_2s_3 + s_1s_2s_4 + s_1s_3s_4$	$(\alpha^3, 0)$
F_{14}	$s_0s_1s_3 + s_0s_2s_3 + s_0s_2s_4 + s_1s_2s_3$	$(\alpha^{10}, 0)$
F_{15}	$s_0s_1s_3 + s_1s_2s_4 + s_1s_3s_4 + s_2s_3s_4$	$(\alpha^{17}, 0)$
F_{16}	$s_0s_2s_3 + s_0s_2s_4 + s_0s_3s_4 + s_1s_2s_4 + s_2s_3s_4$	$(\alpha^{24}, 0)$
F_{17}	$s_0s_1s_3 + s_0s_1s_4 + s_0s_2s_3 + s_0s_2s_4$	$(1, 0)$
F_{18}	$s_0s_1s_2 + s_0s_1s_4 + s_1s_2s_3 + s_1s_2s_4$	$(\alpha^7, 0)$
F_{19}	$s_0s_1s_2 + s_0s_2s_3 + s_2s_3s_4$	$(\alpha^{14}, 0)$
F_{20}	$s_0s_3s_4 + s_1s_2s_3 + s_1s_3s_4$	$(\alpha^{21}, 0)$
F_{21}	$s_0s_1s_4 + s_0s_2s_4 + s_1s_3s_4$	$(\alpha^{28}, 0)$
F_{22}	$s_0s_1s_2 + s_0s_1s_3 + s_0s_2s_4 + s_1s_2s_3 + s_2s_3s_4$	$(\alpha^4, 0)$
F_{23}	$s_0s_1s_3 + s_0s_3s_4 + s_1s_2s_3 + s_1s_2s_4 + s_2s_3s_4$	$(\alpha^{11}, 0)$
F_{24}	$s_0s_1s_4 + s_0s_2s_3 + s_0s_3s_4 + s_1s_2s_4 + s_1s_3s_4 + s_2s_3s_4$	$(\alpha^{18}, 0)$
F_{25}	$s_0s_1s_2 + s_0s_1s_4 + s_0s_2s_3 + s_0s_2s_4 + s_0s_3s_4 + s_1s_2s_3 + s_2s_3s_4$	$(\alpha^{25}, 0)$
F_{26}	$s_0s_1s_2 + s_0s_1s_3 + s_0s_1s_4 + s_0s_3s_4 + s_2s_3s_4$	$(\alpha, 0)$
F_{27}	$s_0s_1s_2 + s_0s_1s_4 + s_0s_3s_4 + s_1s_2s_4 + s_1s_3s_4$	$(\alpha^8, 0)$
F_{28}	$s_0s_1s_2 + s_0s_1s_4 + s_0s_2s_3 + s_0s_2s_4 + s_1s_3s_4 + s_2s_3s_4$	$(\alpha^{15}, 0)$
F_{29}	$s_0s_1s_2 + s_0s_1s_3 + s_0s_2s_4 + s_0s_3s_4 + s_1s_3s_4 + s_2s_3s_4$	$(\alpha^{22}, 0)$
F_{30}	$s_0s_1s_3 + s_0s_1s_4 + s_0s_2s_4 + s_0s_3s_4 + s_1s_2s_3 + s_1s_2s_4 + s_1s_3s_4 + s_2s_3s_4$	$(\alpha^{29}, 0)$

- Shifting of (C_7^{01}, C_{11}^{01}) through its equivalence class until $C_7^{01} = C_7^0 = \alpha^{20}$, that is $(\alpha^5, 0)$ is shifted up to $(\alpha^{20}, 0)$.
- Sum $(\alpha^{20}, 0) + (\alpha^{20}, \alpha^{13}) = (0, \alpha^{13})$, so that $(0, d_2) = (0, \alpha^{13})$.

OUTPUT: Two basic nonlinear filters expressed in their 2-tuple and ANF representations.

1. The 2-tuple $(d_1, 0) = (\alpha^5, 0)$ and its ANF representation $s_0s_1s_2 + s_0s_1s_3 + s_0s_1s_4 + s_0s_2s_3 + s_0s_2s_4 + s_0s_3s_4 + s_1s_2s_3 + s_1s_2s_4 + s_1s_3s_4$.
2. The 2-tuple $(0, d_2) = (0, \alpha^{13})$ and its ANF representation $s_0s_2s_4 + s_0s_3s_4 + s_1s_2s_4$.

Basic filters $(d_1, 0)$ and $(0, d_2)$ range in their corresponding equivalence class (with $2^5 - 1$ filters per class) as it is shown in Tables 1 and 2, respectively. Filter $(d_1, 0)$ includes a unique coset of weight 3 that is (coset 7) as so does $(0, d_2)$ with

Table 2 Class of nonlinear filters $(0, d_2)$ with coset 11 exclusively

Filter	Algebraic normal form	Coeff.
F_0	$s_0s_2s_4 + s_0s_3s_4 + s_1s_2s_4$	$(0, \alpha^{13})$
F_1	$s_0s_1s_3 + s_0s_1s_4 + s_0s_2s_3 + s_1s_3s_4$	$(0, \alpha^{24})$
F_2	$s_0s_1s_2 + s_0s_2s_4 + s_1s_2s_3 + s_1s_2s_4 + s_1s_3s_4 + s_2s_3s_4$	$(0, \alpha^{4})$
F_3	$s_0s_1s_3 + s_0s_2s_3 + s_0s_2s_4 + s_0s_3s_4 + s_1s_2s_3$	$(0, \alpha^{15})$
F_4	$s_0s_1s_3 + s_0s_1s_4 + s_1s_2s_4 + s_2s_3s_4$	$(0, \alpha^{26})$
F_5	$s_0s_1s_2 + s_0s_2s_3 + s_0s_3s_4 + s_1s_2s_3 + s_1s_2s_4$	$(0, \alpha^{6})$
F_6	$s_0s_1s_4 + s_0s_2s_3 + s_1s_2s_3 + s_2s_3s_4$	$(0, \alpha^{17})$
F_7	$s_0s_1s_2 + s_0s_3s_4 + s_1s_2s_3 + s_1s_3s_4 + s_2s_3s_4$	$(0, \alpha^{28})$
F_8	$s_0s_1s_4 + s_0s_2s_4 + s_0s_3s_4 + s_1s_2s_3 + s_1s_3s_4$	$(0, \alpha^{8})$
F_9	$s_0s_1s_2 + s_0s_1s_3 + s_0s_1s_4 + s_0s_2s_4 + s_1s_2s_3 + s_1s_3s_4$	$(0, \alpha^{19})$
F_{10}	$s_0s_1s_2 + s_0s_1s_3 + s_0s_2s_4 + s_1s_2s_4$	$(0, \alpha^{30})$
F_{11}	$s_0s_1s_3 + s_0s_2s_3 + s_1s_2s_3 + s_1s_2s_4$	$(0, \alpha^{10})$
F_{12}	$s_0s_2s_3 + s_1s_2s_4 + s_1s_3s_4 + s_2s_3s_4$	$(0, \alpha^{21})$
F_{13}	$s_0s_2s_3 + s_0s_2s_4 + s_0s_3s_4 + s_1s_3s_4 + s_2s_3s_4$	$(0, \alpha)$
F_{14}	$s_0s_1s_3 + s_0s_1s_4 + s_0s_2s_4 + s_0s_3s_4 + s_2s_3s_4$	$(0, \alpha^{12})$
F_{15}	$s_0s_1s_2 + s_0s_1s_3 + s_0s_1s_4 + s_0s_3s_4 + s_1s_2s_3 + s_1s_2s_4 + s_1s_3s_4$	$(0, \alpha^{23})$
F_{16}	$s_0s_1s_2 + s_0s_1s_4 + s_0s_2s_3 + s_0s_2s_4 + s_1s_2s_4 + s_1s_3s_4$	$(0, \alpha^{3})$
F_{17}	$s_0s_1s_2 + s_0s_1s_3 + s_0s_2s_3 + s_0s_2s_4 + s_1s_3s_4 + s_2s_3s_4$	$(0, \alpha^{14})$
F_{18}	$s_0s_1s_3 + s_0s_2s_4 + s_0s_3s_4 + s_1s_2s_3 + s_1s_2s_4 + s_1s_3s_4 + s_2s_3s_4$	$(0, \alpha^{25})$
F_{19}	$s_0s_1s_3 + s_0s_1s_4 + s_0s_2s_3 + s_0s_2s_4 + s_0s_3s_4 + s_1s_2s_4 + s_1s_3s_4$	$(0, \alpha^{5})$
F_{20}	$s_0s_1s_2 + s_0s_1s_3 + s_0s_1s_4 + s_0s_2s_3 + s_0s_2s_4 + s_1s_2s_3 + s_1s_2s_4 + s_2s_3s_4$	$(0, \alpha^{16})$
F_{21}	$s_0s_1s_2 + s_0s_1s_3 + s_0s_2s_3 + s_0s_3s_4 + s_1s_2s_4 + s_1s_3s_4 + s_2s_3s_4$	$(0, \alpha^{27})$
F_{22}	$s_0s_1s_4 + s_0s_2s_3 + s_0s_2s_4 + s_0s_3s_4 + s_1s_2s_3 + s_1s_2s_4 + s_2s_3s_4$	$(0, \alpha^{7})$
F_{23}	$s_0s_1s_2 + s_0s_1s_3 + s_0s_1s_4 + s_0s_2s_3 + s_0s_3s_4 + s_1s_2s_3 + s_2s_3s_4$	$(0, \alpha^{18})$
F_{24}	$s_0s_1s_2 + s_0s_1s_4 + s_0s_3s_4 + s_1s_2s_4 + s_2s_3s_4$	$(0, \alpha^{29})$
F_{25}	$s_0s_1s_2 + s_0s_1s_4 + s_0s_2s_3 + s_0s_3s_4 + s_1s_3s_4$	$(0, \alpha^{9})$
F_{26}	$s_0s_1s_2 + s_0s_1s_4 + s_0s_2s_4 + s_2s_3s_4$	$(0, \alpha^{20})$
F_{27}	$s_0s_1s_2 + s_0s_1s_3 + s_0s_3s_4$	$(0, 1)$
F_{28}	$s_0s_1s_4 + s_1s_2s_3 + s_1s_2s_4 + s_1s_3s_4$	$(0, \alpha^{11})$
F_{29}	$s_0s_1s_2 + s_0s_2s_3 + s_0s_2s_4 + s_1s_2s_3$	$(0, \alpha^{22})$
F_{30}	$s_0s_1s_3 + s_1s_2s_3 + s_1s_3s_4 + s_2s_3s_4$	$(0, \alpha^{2})$

(coset 11). None of the filters depicted in the previous tables attains the lower bound $LC \geq \binom{5}{3}$ corresponding to the cosets of weight 3. Nevertheless, summing up each one of the ANF representations in Table 1 with every one of the ANF representations in Table 2, we get the 31×31 possible combinations of terms of order 3 that guarantee the cosets of weight 3 (coset 7 and coset 11). Next, the addition of terms of order <3 in ANF representation permits us the generation of all the nonlinear filters of order 3 applied to the previous LFSR that guarantee a linear complexity $LC \geq \binom{5}{3}$.

5 Conclusion

In this work, different representations of nonlinearly filtering functions (ANF, sequential decomposition, N-tuple representation) have been considered and analyzed. At the same time, a method of computing all the nonlinear dynamical filters applied to a LFSR that guarantee the cosets of weight k has been developed. The procedure is based on algebraic operations (addition and shifting operations) on nonlinear filters in different equivalence classes. Starting from a nonlinear filter that is the product of equidistant phases, the computation method formally completes the class of filters with a guaranteed linear complexity of value $LC \geq \binom{L}{k}$. In cryptographic terms, this procedure means an easy way of designing keystream generators for stream ciphers.

Acknowledgments This work has been partially funded by both Ministerio de Economía y Competitividad of Spain under grant number TIN2014-55325-C2-1-R (ProCriCiS), and Comunidad de Madrid, Spain, under grant number S2013/ICE-3095-CM (CIBERDINE).

References

1. Cardell, S.D., Fúster-Sabater, A.: Linear models for the self-shrinking generator based on CA. Journal of Cellular Automata **11**(2–3), 195–211 (2016)
2. Fúster-Sabater, A.: Computing Classes of Cryptographic Sequence Generators. Procedia Computer Science **18**, 2440–2443 (2013)
3. Golomb, S.: Shift-Register Sequences, Revised edn. Aegean Park Press, Laguna Hills, California (1982)
4. Lidl, R., Niederreiter, H.: Finite Fields. In: Enciclopedia of Mathematics and Its Applications vol. 20, Second Ed., Cambridge University Press, Cambridge, UK (1997)
5. Limniotis, K., Kolokotronis, N., Kalouptsidis, N.: On the Linear Complexity of Sequences Obtained by State Space Generators. IEEE Trans. Inform. Theory. **54**, 1786–1793 (2008)
6. Marsaglia, A: Test of DIEHARD http://stat.fsu.edu/geo/diehard.html (1998)
7. Marsaglia, A., Tsang, W.W.: TUFTest, Some difficult-to-pass tests of randomness (2003) http://www.jstatsoft.org/v07/i03/
8. Massey, J.L.: Shift-Register Synthesis and BCH Decoding. IEEE Trans. Information Theory **15**(1), 122–127 (1969)
9. Menezes, A.J., van Oorschot, P., Vanstone, S.: Handbook of Applied Cryptography. CRC Press, New York (1997)

10. Nagaraj, N.: One-Time Pad as a nonlinear dynamical system. Communications in Nonlinear Science and Numerical Simulation **17**, 4029–4036 (2012)
11. Paar, C., Pelzl, J.: Understanding Cryptography. Springer, Berlin Heidelberg (2010)
12. Paul, G., Maitra, S.: RC4 Stream Cipher and Its Variants. CRC Press, Taylor and Francis Group, Boca Raton, Discrete Mathematics and Its Applications (2012)
13. Peinado, A., Fúster-Sabater, A.: Generation of pseudorandom binary sequences by means of LFSRs with dynamic feedback. Mathematical and Computer Modelling **75**(11–12), 2596–2604 (2013)
14. Peinado, A., Munilla, J., Fúster-Sabater, A.: EPCGen2 Pseudorandom Number Generators: Analysis of J3Gen. Sensors **14**, 6500–6515 (2014)
15. Peinado, A., Munilla, J., Fúster-Sabater, A.: Improving the Period and Linear Span of the Sequences Generated by DLFSRs. Advances in Intelligent Systems and Computing, Proc. 7th International Conference on Computational Intelligence in Security for Information Systems, CISIS'14, vol. 299, pp. 397–406. Springer, Berlin (2014)
16. Robshaw, M., Billiet O. (Eds.): New Stream Cipher Designs: The eSTREAM Finalist. *Lecture Notes in Computer Science*, vol. 4986. Springer, Berlin (2008)
17. Ronjom, S., Cid, C.: Nonlinear Equivalence of Stream Ciphers. In: Proc. of Fast Software Encryption, 17th International Workshop, FSE 2010, Seoul, Korea, *Lecture Notes in Computer Science*, vol. 6147, Springer, pp. 40-54 (2010)
18. Rueppel, R.A.: Analysis and Design of Stream Ciphers. Springer, New York (1986)
19. Yet Another SSL (YASSL), Embedded SSL Library (2012) http://www.yassl.com

Implementation of Cryptographic Algorithms for Elliptic Curves

Víctor Gayoso Martínez, Luis Hernández Encinas and Agustín Martín Muñoz

To our colleague and friend Jaime Muñoz Masqué, so that he had something to "think about" and give us ideas for the future, on the occasion of his 65th birthday

Abstract The development of side-channel and fault injection attacks against the implementation of algorithms used in elliptic curve cryptography (ECC), has pointed out that it is not enough to implement efficient algorithms that are secure from a theoretical point of view. In this sense, it is necessary to design algorithms that do not leak information which could allow an attacker to obtain the used keys, thus making the physical implementations of those algorithms resistent to this kind of attacks. In this work, some of the options to implement the scalar multiplication for elliptic curves are described.

Keywords Cryptographic algorithms · Elliptic curves · Fault injection attacks · Scalar multiplication · Side channel attacks

1 Introduction

The widespread use of portable cryptographic tokens (smartcards, USB devices, etc.), whose computation and storage capacity is limited, in addition to the fact that the

V. Gayoso Martínez · L. Hernández Encinas (✉) · A. Martín Muñoz
Institute of Physical and Information Technologies, Spanish National Research Council, C/ Serrano 144, 28006 Madrid, Spain
e-mail: luis@iec.csic.es

V. Gayoso Martínez
e-mail: victor.gayoso@iec.csic.es

A. Martín Muñoz
e-mail: agustin@iec.csic.es

M. Castrillón López et al. (eds.), *Geometry, Algebra and Applications: From Mechanics to Cryptography*, Springer Proceedings in Mathematics & Statistics 161, DOI 10.1007/978-3-319-32085-4_11

physical implementations of cryptographic algorithms in those devices are specially threatened by side-channel or fault injection attacks (commonly known as physical attacks), highlights the importance of analyzing the characteristics of the elliptic curve cryptography (ECC) algorithms used by those devices.

Scalar multiplication is the basic operation in ECC when physical devices with limited computational and memory resources are used. Thus, it is necessary to take into account the cost of its implementation in terms of the type and number of operations to perform. As it is well known, a scalar multiplication in additive groups is algorithmically analogous to an exponentiation in multiplicative groups. There are, however, some differences that justify the study of scalar multiplication on its own. For instance, the fact that a point inversion within the group of points on an elliptic curve has a relatively low cost, allows some algorithmic optimizations.

In this work we have focused on elliptic curves, $E(\mathbb{F}_p)$ with infinity point $\mathcal{O}$, defined over prime fields, $\mathbb{F}_p$, $p > 3$, because they are the most widely used in practical applications (the American National Security Agency has selected the option of prime fields for its Suite B, [18]), and because the use of ECC over binary fields is protected by several patents.

The rest of the work is organized as follows: in Sect. 2 some concepts related to elliptic curve arithmetics are presented. Section 3 describes the classical algorithms for scalar multiplication. Finally, algorithms that include some type of countermeasure to avoid some physical attacks are analyzed in Sect. 4. For a more exhaustive treatment of ECC and its arithmetics, the interested reader can see [5, 9].

2 Elliptic Curve Arithmetics

The computational cost of additions and subtractions is usually considered negligible as compared to that of multiplications. In spite of this consideration being asymptotically correct for personal computers and similar devices, this is not accurate in the case of implementing algorithms in small embedded cryptographic devices which use an arithmetic coprocessor capable to add, subtract, multiply, and make modular operations such as multiplication or squaring.

In most cases, modular additions and subtractions are carried out by means of normal additions and subtractions and, respectively, conditional subtractions and additions. In practice, these conditional operations imply a side-channel vulnerability which could be avoided by performing the operations unconditionally, that is, performing dummy operations half of the time. On the other hand, as each operation carried out by the coprocessor requires an extra software processing to configure and start it, it seems clear that the cost of additions and subtractions is not negligible in practice.

However, the most computationally costly operation is multiplication. It is generally considered that, for operands of a few hundred of bits, $S/M \sim 0.8$, M being the cost of a modular multiplication and S that of a squaring. The cost of a modular inversion is much higher than that of a multiplication; in smartcards it has been observed that $I/M \sim 100$, I being the cost of a modular inversion [21].

Most of the computation time employed in the aforementioned embedded devices is related to the curve point arithmetic, namely additions, subtractions, multiplications by scalars and inversions. The remaining operations (conditional branching, assignments, loop processing, etc.), have a negligible cost as compared to the computational cost of the above arithmetic operations.

In some cases, it is adequate to use Jacobian projective coordinates [9], or modified (or mixed) Jacobian coordinates [2], to avoid performing some operations. For example, if two points in Jacobian coordinates have the same Z coordinate, they can be added very efficiently [15]. Indeed, if $P_1 = (X_1 : Y_1 : Z)$ and $P_2 = (X_2 : Y_2 : Z)$, with $X_1 \neq X_2$, the sum $P_1 + P_2 = P_3$, with $P_3 = (X_3 : Y_3 : Z_3)$ can be computed as:

$$\begin{cases} X_3 = D - B - C, \\ Y_3 = (Y_2 - Y_1)(B - X_3) - Y_1(C - B), \\ Z_3 = Z(X_2 - X_1), \end{cases} \text{with} \begin{cases} A = (X_2 - X_1)^2, \\ B = X_1 A, \\ C = X_2 A, \\ D = (Y_2 - Y_1)^2. \end{cases}$$

This Meloni's addition formula [15] is known as co-Z addition, and it has a very small computational cost. An interesting property of this formula is that the point P_1 can be converted after the addition into another representative $P_1' = (X_1' : Y_1' : Z_3)$, with no cost. It can be seen that $Z_3 = Z(X_2 - X_1)$, which yields $X_1' = X_1(X_2 - X_1)^2 = X_1 A$ and $Y_1' = Y_1(X_2 - X_1)^3 = Y_1(C - B)$, expressions which are part of the Meloni's addition formula above.

Co-Z addition and update (ZADDU) uses Meloni's addition formula, such that $\mathsf{ZADDU}(P_1, P_2) = (P_3, P_1')$. Goundar et al. [7] extended this operation to a conjugate addition, referred to as $\mathsf{ZADDC}(P_1, P_2) = (P_3, P_4)$, observing that the previous co-Z addition can also yield $P_1 - P_2 = (X_4 : Y_4 : Z_3)$ with [21]

$$\begin{cases} X_4 = (Y_2 + Y_1)^2 - B - C, \\ Y_4 = (Y_2 + Y_1)(B - X_4) - Y_1(C - B). \end{cases}$$

3 Efficient Scalar Multiplication Algorithms

In this section, the most efficient classical algorithms to compute the scalar multiplication when there is no threat of potential side-channel attacks are presented (assuming, for instance, that the calculations to be performed are public, or that they take place in a secure environment).

The simpler binary methods are commonly called *double-and-add* algorithms, either left-to-right or right-to-left. There are also methods based on the *non-adjacent form* (NAF) representation of the scalar with sign, which use can improve the efficiency of the simple binary methods. Moreover, the *sliding window* methods are also presented. These are methods that use time-memory trade-offs to speed up the scalar multiplication, provided that extra memory is available.

In the following, when the computational cost of an algorithm is mentioned, it is assumed that the curve point is represented in affine coordinates. This allows several improvements when using algorithms which operate with the scalar bits from left to right. This assumption is reasonable because the use of affine coordinates is a standard procedure in most protocols.

3.1 Simple Binary Algorithms

There are several formulations to calculate the scalar multiplication (kP) starting from the binary decomposition of the scalar, $k = (k_{l-1}k_{l-2}\dots k_0)_2$. If, for example, the following formulas are considered,

$$kP = k_0P + k_1 2P + \cdots k_{l-1}2^{l-1}P,$$
$$kP = k_0P + 2(k_1P + 2(\cdots + 2(k_{l-1}P)\cdots)),$$

the algorithms to perform the corresponding operations are known as double-and-add algorithms (due to the additive group structure), and are similar to the classical exponentiation algorithms based on squaring and multiplying. Algorithm 1 is called *left-to-right* because it starts using the scalar bits from the most significant to the least significant. In Algorithm 2 the scanning of the bits is made in the opposite direction and, thus, it is called right-to-left algorithm.

Algorithm 1 Double-and-add left-to-right scalar multiplication

Input: $P \in E(\mathbb{F}_p)$, $k = (k_{l-1}k_{l-2}\dots k_0)_2$
Output: kP
Uses: P, Q
$Q \leftarrow \mathcal{O}$
for $i = l-1$ to 0 **do**
  $Q \leftarrow 2Q$
  if $k_i = 1$ **then**
    $Q \leftarrow Q + P$
  end if
end for
return Q

Both algorithms have the same complexity in terms of point operations. However, their cost differ when considering field multiplications. In Algorithm 1 it is possible to use the mixed affine-projective addition formula when the affine coordinates of the input point are known. Furthermore, in [12] it was pointed out that Algorithm 2 allows using the Jacobian addition formula at step 8 and the modified Jacobian doubling formula at step 10, provided the Jacobian coordinates of point Q and the modified Jacobian coordinates of point R are known. This allows fast point multiplication on elliptic curves without precomputation.

Algorithm 2 Double-and-add right-to-left scalar multiplication

```
Input: P ∈ E(F_p), k = (k_{l-1} k_{l-2} ... k_0)_2
Output: kP
Uses: P, Q, R
Q ← O
R ← P
for i = 0 to l − 1 do
    if k_i = 1 then
        Q ← Q + R
    end if
    R ← 2R
end for
return Q
```

3.2 Algorithms with Non-adjacent Forms (NAF)

It is relevant to notice that an inversion operation $P \leftarrow -P$ has a computational cost almost negligible. Thus, it is interesting to use signed representations to decrease the number of additions to be performed in the scalar multiplication. This can be made by first converting the scalar into its non-adjacent form, which is a binary representation with sign in such a way that there is no two consecutive bits equal to 1. That is, the NAF representation of an integer $k \in \mathbb{N}^*$ is $k = (k_{l-1}k_{l-2}\ldots k_0)_{\mathsf{NAF}}$ with $k_i \in \{-1, 0, 1\}$, $0 \leq i < l-1$ and $k_{l-1} = 1$ such that, for any consecutive digits k_i and k_{i+1}, $k_i k_{i+1} = 0$.

There is a unique NAF representation of a given scalar so that its length is the same than the binary representation of the scalar or has one more digit, its number of non-zero digits is always minimal among base 2 signed representations for a given scalar, and there exists a simple algorithm to obtain the NAF representation of a positive integer which uses only low-cost operations (see Algorithm 3).

Algorithm 3 Calculation of the NAF representation of an scalar

```
Input: k ∈ N*
Output: k_NAF
i ← 0
while k ≥ 1 do
    if k (mod 2) = 1 then
        k_i ← 2 − (k (mod 4))
        k ← k − k_i
    else
        k_i ← 0
    end if
    k ← k/2
    i ← i + 1
end while
return k_NAF = (k_{l-1} k_{l-2} ... k_0)_NAF
```

Algorithm 4 shows how to calculate the left-to-right scalar multiplication using the NAF decomposition of the scalar. Algorithm 5 [12] includes the computation of the NAF representation of the scalar and is a right-to-left variant of Algorithm 4.

Algorithm 4 Left-to-right NAF scalar multiplication

Input: $P \in E(\mathbb{F}_p)$, $k = (k_{l-1}k_{l-2}\ldots k_0)_{\mathsf{NAF}}$
Output: kP
Uses: P, Q
$Q \leftarrow \mathcal{O}$
for $i = l - 1$ to 0 **do**
  $Q \leftarrow 2Q$
  if $k_i = 1$ **then**
    $Q \leftarrow Q + P$
  end if
  if $k_i = -1$ **then**
    $Q \leftarrow Q + (-P)$
  end if
end for
return Q

3.3 Window NAF Algorithms

If some odd multiples of the input point are precomputed, a scalar multiplication can be calculated in a more efficient way. The basic idea is to operate with the bits of the scalar in windows of an adequate size and store in a table the result of multiplying the point by each window of bits.

As an example, if the scalar has two bits 11, the calculation which is made according to the left-to-right double-and-add algorithm is $2(2Q + P) + P$. However, if $3P$ is known, the same result is obtained by computing $2(2Q) + 3P$, thus saving an addition. This idea can be extended to blocks with more bits (windows of different sizes) and NAF representations [6, 14]. With right-to-left algorithms, a similar strategy can be used. In this case some values which are stored during intermediate calculations of the scalar multiplication are combined at the end [16, 22].

In [9], two families of algorithms which use blocks of several bits (windows) at the same time are defined, namely, sliding window NAF, and window width-w NAF algorithms. Both can be implemented either left-to-right or right-to-left. The algorithm to perform a left-to-right sliding window NAF scalar multiplication is shown in Algorithm 6.

Algorithm 7 shows a right-to-left window width-w NAF. In this case, the width-w non-adjacent form of the scalar is computed on-the-fly.

Algorithm 5 Right-to-left NAF scalar multiplication

```
Input: P ∈ E(F_p), k ∈ N*
Output: kP
Uses: P, R
Q ← O
R ← P
while k ≥ 1 do
  if k (mod 2) = 1 then
    u ← 2 − (k (mod 4))
    k ← k − u
    if u = 1 then
      Q ← Q + R
    else
      Q ← Q + (−R)
    end if
  end if
  k ← k/2
  R ← 2R
end while
return Q
```

Algorithm 6 requires to store $(2^w - (-1)^w)/3 - 1$ more points than Algorithm 4. In the case of Algorithm 7, $2^{w-1} - 1$ more points need to be stored, as compared to Algorithm 5.

Algorithms 6 and 7 have a similar efficiency. Choosing the optimal one depends on the length of the scalar, and also on the amount of available memory (the larger the memory, the larger the possible width of the window).

These NAF techniques were generalized in [16] with the signed fractional window representation, which allows to use any set of consecutive odd digits $\pm 1, \pm 3, \pm 5, \ldots$. In this way the efficiency of the scalar multiplication can be optimized by using a window of width $w \geq 4$ and all the available memory storage [21].

4 Algorithms Which Include Countermeasures

As mentioned in Sect. 2, side-channel attacks use some assignments or conditional branching of the algorithms, which frequently depend on the bits of the key, to obtain information about the latter. Thus, a large variety of countermeasures have been proposed in order to modify the algorithms such that their execution includes a series of regular operations which are independent of the value of the scalar bits. Those are known as *regular* algorithms, and the sequence of operations they perform is constant no matter whether the bits of the scalar are 0 or 1. In this section, several ways of applying this regularity are described.

Algorithm 6 Left-to-right sliding window NAF scalar multiplication

```
Input: P ∈ E(F_p), k = (k_{l-1} k_{l-2} ... k_0)_NAF, w ≥ 2
Output: kP
Uses: Q, P_1, P_3, ..., P_m, where m = 2^((2^w - (-1)^w)/3) - 1
Q ← 𝒪
i ← l - 1
Precomputations
for i = 1 to m by 2 do
    P_i ← iP
end for
Main loop
while i ≥ 0 do
    if k_i = 0 then
        Q ← 2Q
        i ← i - 1
    else
        s ← max(i - w + 1, 0)
        while k_s = 0 do
            s ← s + 1
        end while
        u ← (k_i ... k_s)_NAF
        for j = 1 to i - s + 1 do
            Q ← 2Q
        end for
        if u > 0 then
            Q ← Q + P_u
        end if
        if u < 0 then
            Q ← Q + (-P_{-u})
        end if
        i ← s - 1
    end if
end while
return Q
```

4.1 Double-and-Add Always

In Sect. 3.1 it can be observed that in step 8 of Algorithm 1 and Algorithm 2 a point addition is performed only when the value of the processed scalar bit is 1. This can leak information to an attacker. An obvious countermeasure against side-channel attacks, first proposed in [3], is to always perform a point addition, no matter if $k_i = 1$ or $k_i = 0$ (in this case the point addition is a dummy operation). The method is thus called *double-and-add always*. Algorithm 8 presents the right-to-left variant of this countermeasure, where a dummy addition is included in step 10.

It is possible that applying a countermeasure against a certain kind of attack could generate a vulnerability to other attacks. This is the case of Algorithm 8, because if an attacker is able to induce a fault during the addition of points (this can be made in steps 8 or 10), it is possible to deduce the value of the corresponding bit

Algorithm 7 Right-to-left on-the-fly window width-w NAF scalar multiplication

Input: $P \in E(\mathbb{F}_p)$, $k = (k_{l-1}k_{l-2}\dots k_0)_2$, $w \geq 2$
Output: kP
Uses: R, $Q_1, Q_3, \dots, Q_m$, where $m = 2^{w-1} - 1$
$R \leftarrow P$
$Q_1, Q_3, \dots, Q_m \leftarrow \mathcal{O}$
Main loop
while $k \geq 1$ **do**
  if $k \pmod 2 = 1$ **then**
    $t \leftarrow k \pmod{2^w}$ $\rightsquigarrow$ where $k \pmod{2^w} \in [-2^{w-1}, 2^{w-1} - 1]$
    if $t > 0$ **then**
      $Q_t \leftarrow Q_t + R$
    end if
    if $t < 0$ **then**
      $Q_{-t} \leftarrow Q_{-t} - R$
    end if
    $k \leftarrow k - t$
  end if
  $R \leftarrow 2R$
  $k \leftarrow k/2$
end while
Postcomputations
for $i = 3$ to m by 2 **do**
  $Q_1 \leftarrow Q_1 + iQ_i$
end for
return Q_1

Algorithm 8 Right-to-left double-and-add-always scalar multiplication

Input: $P \in E(\mathbb{F}_p)$, $k = (k_{l-1}k_{l-2}\dots k_0)_2$
Output: kP
Uses: Q, R, T
$Q, T \leftarrow \mathcal{O}$
$R \leftarrow P$
for $i = 0$ to $l - 1$ **do**
  if $k_i = 1$ **then**
    $Q \leftarrow Q + R$
  else
    $T \leftarrow T + R$
  end if
  $R \leftarrow 2R$
end for
return Q

by just checking the output of the algorithm. Indeed, a correct result implies that the operation that was performed was dummy, and the bit is $k_i = 0$. On the contrary, a wrong result indicates that $k_i = 1$. This is the basic idea used in the *safe-error* attacks, introduced in [23]. The positive aspect, however, is that when using Algorithm 8 it is easy to check whether there has been an attempt to induce a fault because, as at the end of the algorithm it is expected that $Q = kP$, $T = (2^l - k - 1)P$ and $R = 2^l P$, it would be enough to check that $Q + T + P = R$ to ensure that no attack has occured.

4.2 Montgomery Ladder

Another well known countermeasure is the *Montgomery ladder*, presented in Algorithm 9. It was introduced in [17] to accelerate the scalar multiplication on a specific class of curves, and has been generalized to elliptic curves over fields of large characteristic [1, 4, 10].

Algorithm 9 allows to accelerate the scalar multiplication provided that the y coordinate of the result is not needed. This occurs in ECDSA and many other cryptographic protocols. As it can be observed, there is no dummy operation in the Montgomery ladder algorithm and, thus, it is not threatened by safe-error attacks. Another advantage is that it can be easily parallelized [4].

Algorithm 9 Montgomery ladder scalar multiplication

1: *Input:* $P \in E(\mathbb{F}_p)$, $k = (k_{l-1}k_{l-2}\ldots k_0)_2$ where $k_{l-1} = 1$
2: *Output:* kP
3: *Uses:* Q_0, Q_1
4: $Q_0 \leftarrow P$
5: $Q_1 \leftarrow 2P$
6: **for** $i = l - 2$ to 0 **do**
7: $\quad Q_{1-k_i} \leftarrow Q_0 + Q_1$
8: $\quad Q_{k_i} \leftarrow 2Q_{k_i}$
9: **end for**
10: **return** Q_0

4.3 Joye Double and Add Ladder

As a general rule, in order to prevent or avoid side-channel attacks, right-to-left algorithms are usually better than left-to-right ones. In [11], the author proposed a right-to-left algorithm (see Algorithm 10) which, besides, is not affected by safe-error attacks, a powerful type of fault induction attacks. Joye's algorithm, similarly to Algorithm 9, carries out a doubling and an addition for every bit of the scalar.

Algorithm 10 Joye double-and-add ladder scalar multiplication

Input: $P \in E(\mathbb{F}_p)$ y $k = (k_{l-1}k_{l-2}\dots k_0)_2$
Output: kP
Uses: Q_0, Q_1
$Q_0 \leftarrow \mathcal{O}$
$Q_1 \leftarrow P$
for $i = 0$ to $l - 1$ **do**
    $Q_{1-k_i} \leftarrow 2Q_{1-k_i} + Q_{k_i}$
end for
return Q_0

4.4 *Joye m-ary Ladders*

Algorithm 11, called regular m-ary, was presented in [13] and, with $m = 2$, is very similar to the Montgomery ladder. Regarding its performance, Algorithm 9, which uses only the x coordinate, is the fastest one. The performance of Algorithm 11 with $m = 2$ is better than that of Algorithm 10.

Algorithm 11 Regular left-to-right Joye m-ary scalar multiplication

Input: $P \in E(\mathbb{F}_p)$, $k = (k_{l_m-1}k_{l-m-2}\dots k_0)_m$ where $k > 0$
Output: kP
Uses: $Q, R_0, R_1, \dots, R_{m-1}$
$Q \leftarrow -P$
Precomputations
for $i = 0$ to $m - 1$ **do**
    $R_i \leftarrow (m + i - 1)P$
end for
Main loop
for $i = l_m - 1$ to 0 **do**
    $Q \leftarrow mQ + R_{k_i}$ $\rightsquigarrow$ t duplications and an addition if $m = 2^t$
end for
Final correction
$Q \leftarrow Q + P$
return Q

4.5 *Co-Z Ladders*

New scalar multiplication methods based on the co-Z Jacobian arithmetics ($\mathcal{J}$) were presented in [7, 20], later on extended in [8, 19]. As an example, the Montgomery ladder algorithm can be converted to Algorithm 12. In this way it is possible to take advantage of the ZADDU and ZADDC co-Z Jacobian addition formulas.

In [15] it was pointed out that a formula of ZADDU which only uses the $(X:Y)$ coordinates could be used in scalar multiplication algorithms (in this case the formula is denoted ZADDU'). The Z coordinate can be generally obtained at the end with a few extra field operations. This can also be applied to Algorithm 12 (see Algorithm 13), by using the ZADDC' operation, which is a $(X:Y)$-only variant of ZADDC [8, 20].

Algorithm 12 Montgomery ladder scalar multiplication using $\mathcal{J}$ with co-Z addition

Input: $P=(x,y)\in E(\mathbb{F}_p)$, $k=(k_{l-1}k_{l-2}\ldots k_0)_2$ where $k_{l-1}=1$
Output: kP
Uses: Q_0, Q_1
$Q_1 \leftarrow (X_{2P}:Y_{2P}:Z_{2P})$ $\rightsquigarrow$ Jacobian representation of $2P$
$Q_0 \leftarrow (xZ_{2P}^2:yZ_{2P}^3:Z_{2P})$ $\rightsquigarrow$ Jacobian representation of P
for $i=l-2$ to 0 **do**
  $(Q_{1-k_i},Q_{k_i}) \leftarrow$ ZADDC(Q_{k_i},Q_{1-k_i})
  $(Q_{k_i},Q_{1-k_i}) \leftarrow$ ZADDU(Q_{1-k_i},Q_{k_i})
end for
return Q_0

Algorithm 13 Montgomery ladder scalar multiplication using $\mathcal{J}$ with $(X:Y)$-only co-Z addition

Input: $P=(x,y)\in E(\mathbb{F}_p)$, $k=(k_{l-1}k_{l-2}\ldots k_0)_2$ where $k_{l-1}=1$
Output: kP
Uses: Q_0, Q_1
$Q_1 \leftarrow (X_{2P}:Y_{2P})$
$Q_0 \leftarrow (xZ_{2P}^2:yZ_{2P}^3)$
for $i=l-2$ to 1 **do**
  $(Q_{1-k_i},Q_{k_i}) \leftarrow$ ZADDC'(Q_{k_i},Q_{1-k_i})
  $(Q_{k_i},Q_{1-k_i}) \leftarrow$ ZADDU'(Q_{1-k_i},Q_{1-k_i})
end for
$(Q_{1-k_0},Q_{k_0}) \leftarrow$ ZADDC'(Q_{k_0},Q_{1-k_0})
$Z \leftarrow xY_{Q_{k_0}}(X_{Q_0}-X_{Q_1})$
$\lambda \leftarrow yX_{Q_{k_0}}$
$(Q_{k_0},Q_{1-k_0}) \leftarrow$ ZADDU'(Q_{1-k_0},Q_{k_0})
return $(\lambda^2 X_{Q_0}:\lambda^3 Y_{Q_0}:Z)$

Acknowledgments This work has been partially supported by project TIN2014-55325-C2-1-R (ProCriCiS), funded by Ministerio de Economía y Competitividad, Spain.
We thank Jaime Muñoz Masqué for his continuous and challenging ideas to address new research objectives, as well as for all his useful suggestions to overcome so many obstacles.

References

1. Brier, E., Joye, M.: Weierstrass elliptic curves and side-channel attacks. Lect. Notes Comput. Sci. **2274**, 335–345 (2002)
2. Cohen, H., Miyaji, A., Ono, T.: Efficient elliptic curve exponentiation using mixed coordinates. Lect. Notes Comput. Sci. **1514**, 51–65 (1998)
3. Coron, J.S.: Resistance against differential power analysis for elliptic curve cryptosystems. Lect. Notes Comput. Sci. **1717**, 292–302 (1999)
4. Fischer, W., Giraud, C., Knudsen, E.W., Seifert, J.P.: Parallel scalar multiplication on general elliptic curves over $\mathbb{F}_p$ hedged against non-differential side-channel attacks. Cryptology ePrint Archive, Report 2002/007
5. Fúster Sabater, A., Hernández Encinas, L., Martín Muñoz, A., Montoya Vitini, F., Muñoz Masqué, J.: Criptografía, protección de datos y aplicaciones. RA-MA, Madrid (2012)
6. Gordon, D.M.: A survey of fast exponentiation methods. J. Algorithms **27**(1), 129–146 (1998)
7. Goundar, R., Joye, M., Miyaji, A.: Co-Z addition formula and binary ladders on elliptic curves. Lect. Notes Comput. Sci. **6225**, 65–79 (2010)
8. Goundar, R., Joye, M., Miyaji, A., Rivain, A., Venelli, A.: Scalar multiplication on Weierstrass elliptic curves from co-Z arithmetic. J. Cryptogr. Eng. **1**(2), 161–176 (2011)
9. Hankerson, D., Menezes, A.J., Vanstone, S.: Guide to Elliptic Curve Cryptography. Springer, New York (2004)
10. Izu, T., Takagi, T.: A fast parallel elliptic curve multiplication resistant against side channel attacks. Lect. Notes Comput. Sci. **2274**, 280–296 (2002)
11. Joye, M.: Highly regular right-to-left algorithms for scalar multiplication. Lect. Notes Comput. Sci. **4727**, 135–147 (2007)
12. Joye, M.: Fast point multiplication on elliptic curves without precomputation. In: Proceedings of WAIFI 2008
13. Joye, M.: Highly regular m-ary powering ladders. Lect. Notes Comput. Sci. **5867**, 350–363 (2009)
14. Koyama, K., Tsuruoka, Y.: Speeding up elliptic cryptosystems by using a signed binary window method. Lect. Notes Comput. Sci. **740**, 345–357 (1993)
15. Meloni, N.: New point addition formulae for ECC applications. Lect. Notes Comput. Sci. **4547**, 189–201 (2007)
16. Möller, B.: Improved techniques for fast exponentiation. Lect. Notes Comput. Sci. **2587**, 298–312 (2003)
17. Montgomery, P.: Speeding the Pollard and elliptic curve methods of factorization. Math. Comput. **48**, 243–264 (1987)
18. NSA, NSA Suite B cryptography. National Security Agency (2005)
19. Rivain, M.: Fast and regular algorithms for scalar multiplication over elliptic curves. Cryptology ePrint Archive, Report 2011/338
20. Venelli, A., Dassance, F.: Faster side-channel resistant elliptic curve scalar multiplication. Contemp. Math. **521**, 29–40 (2010)
21. Verneuil, V.: Elliptic curve cryptography and security of embedded devices. Ph.D. Thesis. Ècole Doctorale de Mathèmatiques et Informatique, Universitè de Bordeaux (France) (2010)
22. Yao, A.C.C.: On the evaluation of powers. SIAM J. Comput. **5**(1), 100–103 (1976)
23. Yen, S.M., Joye, M.: Checking before output may not be enough against fault-based cryptanalysis. IEEE Trans. Comput. **49**(9), 967–970 (2000)

Supermanifolds, Symplectic Geometry and Curvature

Rosalía Hernández-Amador, Juan Monterde and José Vallejo

Jaime Muñoz-Masqué, maestro y amigo, en su 65 aniversario

Abstract We present a survey of some results and questions related to the notion of scalar curvature in the setting of symplectic supermanifolds.

Keywords Supermanifolds · Symplectic curvature · Scalar curvature

1 Introduction

Supermanifolds appeared in Mathematics as a way to unify the description of bosons and fermions in Physics. Of course, there would be nothing special about them if the resulting theory were just the juxtaposition of separate theorems, what is really interesting is the possibility of new phenomena arising from the interaction of both

---

Dedicated to Muñoz-Masqué, teacher and friend, on his 65th birth anniversary.

R. Hernández-Amador
Departamento de Matemáticas, Universidad de Sonora, 83000 Hermosillo, Sonora, Mexico
e-mail: guadalupehernandez@correoa.uson.mx

J. Monterde
Departament de Geometria i Topologia, Universitat de València, 46100
Burjassot, València, Spain
e-mail: juan.l.monterde@uv.es

J. Vallejo (✉)
Facultad de Ciencias, Universidad Autónoma de San Luis Potosí, 78290
San Luis Potosí, Mexico
e-mail: jvallejo@fc.uaslp.mx

M. Castrillón López et al. (eds.), *Geometry, Algebra and Applications: From Mechanics to Cryptography*, Springer Proceedings in Mathematics & Statistics 161, DOI 10.1007/978-3-319-32085-4_12

(the bosonic and the fermionic) worlds. From the point of view of Physics, the most prominent exponent is the phenomenon of supersymmetry, much questioned these days in view of the absence of experimental evidence coming from the LHC research, but from a purely mathematical point of view there is the exciting possibility of investigating geometric structures which can be understood only by looking at them through "fermionic lenses".

Symplectic scalar curvature is one of these structures: if one starts out with a connection on a usual manifold, it is straightforward to define its associated curvature, but if a refinement such as Ricci or scalar curvature is desired (as in General Relativity), then a non-degenerate bilinear form (a second-order covariant tensor field) is required to take the relevant traces. Riemannian geometry enters the stage when that tensor field is taken symmetric, leading to a plethora of well-known results, but there is another possibility. A symplectic form could be used to make the successive contractions needed to pass from the curvature four-tensor to the scalar curvature, but it is readily discovered that the would-be symplectic scalar curvature obtained this way vanishes due to the different symmetries involved (the Ricci tensor is symmetric and is contracted with the skew-symmetric symplectic form). Thus, it would seem that there is no room for a non-trivial Riemannian-symplectic geometry, an idea further supported from the observation that locally Riemannian and symplectic geometries are quite opposite to each other, as in the symplectic case there are no invariants because of the Darboux theorem.

However, things are different if we allow for supermanifolds. In this case, there are two variants of symplectic forms, even and odd ones, and it is remarkable that, while even symplectic forms lead to the same results as in the non graded setting, for odd symplectic manifolds it is possible, a priori, to define a symplectic scalar curvature, because the symmetries involved in this setting do not forbid its existence. However, the explicit construction of examples is very difficult, and in this paper we try to explain why. The ultimate reason is that the structure of odd symplectic manifolds is very restrictive. In particular, they strongly depend on the existence on an isomorphism between the tangent bundle TM and the Batchelor bundle E (that is, the vector bundle over M such that the supermanifold $(M, \mathscr{A})$ satisfies $\mathscr{A} \simeq \Gamma \Lambda E$). When this isomorphism comes from a non-degenerate bilinear form on TM with definite symmetry (e.g., a Riemannian metric or a symplectic form), the symmetries of the graded Ricci tensor lead to a trivial scalar curvature, as in the non-graded case.

While we will not deepen into the physical applications, neither of this odd symplectic curvature nor supersymplectic forms in general (for this, see [1, 2, 4, 9]), we will offer a detailed review of the mathematics involved in this construction under quite general conditions, avoiding excessive technicalities with the aim of making this topic available to a wider audience.

2 Preliminaries

Let M be a differential manifold, let $\mathscr{X}(M)$ denote the $\mathscr{C}^\infty(M)$-module of its vector fields, and let ∇ be a linear (Koszul) connection on it. The curvature of ∇ is the operator $\mathrm{Curv} : \mathscr{X}(M) \times \mathscr{X}(M) \to \mathrm{End}\mathscr{X}(M)$ such that

$$\mathrm{Curv}(X, Y) = [\nabla_X, \nabla_Y] - \nabla_{[X,Y]},$$

where $[X, Y]$ is the Lie bracket of vector fields and $[\nabla_X, \nabla_Y]$ is the commutator of endomorphisms. Given a Riemannian metric on M (that is, a symmetric, positive-definite, covariant 2-tensor field $g \in S^2_+(M)$), there is a particular linear connection on M, the Levi-Civita connection, such that $\nabla g = 0$. With the aid of the metric, two further contractions of the curvature can be defined, the first one leading to the Ricci covariant 2-tensor

$$\mathrm{Ric}(X, Y) = \mathrm{Tr}_g(Z \to \mathrm{Curv}(X, Z)Y), \tag{1}$$

and the second one to the Riemannian scalar curvature

$$S = \mathrm{Tr}(g^{-1}\mathrm{Ric}). \tag{2}$$

Let us remark that the Ricci tensor (1) is symmetric, as it is g, so the contraction in (2) does not vanish a priori.

Now suppose that we use a compatible symplectic form $\omega \in \Omega^2(M)$ (that is, such that $\nabla\omega = 0$) to compute these contractions. Using a superindex to distinguish them from the previous ones, we obtain

$$\mathrm{Ric}^\omega(X, Y) = \mathrm{Tr}_\omega(Z \to \mathrm{Curv}(X, Z)Y), \tag{3}$$

and

$$S^\omega = \mathrm{Tr}(\omega^{-1}\mathrm{Ric}^\omega). \tag{4}$$

The symplectic Ricci tensor (3) is again symmetric, but this time the contraction in (4) involves the skew-symmetric ω^{-1}, so we get $S^\omega = 0$.

The study of symplectic manifolds (M, ω) endowed with a connection ∇ such that $\nabla\omega = 0$ can be carried on along lines similar to those of Riemannian geometry (see [7]). The resulting Fedosov manifolds appeared first in the deformation quantization of Poisson manifolds (see [5]). The fact that a basic local invariant such as the scalar curvature vanishes on any Fedosov manifold has led to a certain lack of interest in its use in Physics and Mathematics, aside from the mentioned rôle in deformation quantization. However, if supermanifolds are considered a new possibility appears. There are two classes of symplectic forms on a supermanifold and, as we see below, one of them has the symmetry properties required to obtain a non-trivial contraction defining the symplectic scalar curvature.

A supermanifold can be thought of as a non-commutative space of a special kind, one in which the sheaf of commutative rings of $\mathscr{C}^\infty(M)$ functions has been replaced by a sheaf of $\mathbb{Z}_2$-graded supercommutative algebras, that is, to each open subset $U \subset M$ of a manifold, we assign an algebra $\mathscr{A}(U) = \mathscr{A}_0(U) \oplus \mathscr{A}_1(U)$ with a product such that $\mathscr{A}_i(U) \cdot \mathscr{A}_j(U) \subset \mathscr{A}_{(i+j)\mathrm{mod}2}(U)$ and $a \cdot b = (-1)^{|a||b|} b \cdot a$, where $|a|$, $|b|$ denote the $\mathbb{Z}_2$ degree of the elements $a, b \in \mathscr{A}(U)$. An exposition of the basic facts about supermanifolds oriented to physical applications can be found in [15].

For completeness, let us give here the definition: a real supermanifold is a ringed space $(M, \mathscr{A})$, where $\mathscr{A}$ is a sheaf of $\mathbb{Z}_2$-graded commutative $\mathbb{R}$-algebras such that:

(a) If $\mathscr{N}$ denotes the sheaf of nilpotents of $\mathscr{A}$, then $\mathscr{A}/\mathscr{N}$ induces on M the structure of a differential manifold.
(b) The subsheaf $\mathscr{N}/\mathscr{N}^2$ is a locally free sheaf of modules, with $\mathscr{A}$ *locally* isomorphic to the exterior sheaf $\bigwedge\left(\mathscr{N}/\mathscr{N}^2\right)$.

The sheaf of differential forms on a manifold M, where $\Omega(U) = \bigoplus_{p\in\mathbb{Z}} \Omega^p(U)$, provide a good example. The nilpotents in this case are all the $\alpha \in \Omega^p(M)$ with $p \geq 1$, so $\mathscr{A}/\mathscr{N} = \mathscr{C}^\infty(M)$ (the smooth functions on M). Moreover, $\mathscr{N}/\mathscr{N}^2 = \Omega^1(M)$, the space of 1-forms, is locally generated by the differentials $\mathrm{d}x^1, \ldots, \mathrm{d}x^m$ of the functions x^i of a chart on M. Thus, as a model for a supermanifold we can think of a usual manifold M endowed with "superfunctions", which are just differential forms and can be classified as even and odd by their degree. From now on, until otherwise explicitly stated, we will assume that our supermanifold is $(M, \Omega(M))$, and sometimes we will refer to it as the Koszul or Cartan–Koszul supermanifold.[1]

The replacement of $\mathscr{C}^\infty(M)$ by $\Omega(M)$ leads to the definition of other basic structures of differential geometry. For instance, (super) vector fields on the supermanifold $(M, \Omega(M))$ are now the derivations $\mathrm{Der}\,\Omega(M)$ (such as the exterior differential d, which has degree $|\mathrm{d}| = 1$, the Lie derivative $\mathscr{L}_X$, which has degree $|\mathscr{L}_X| = 0$, or the insertion i_X, which has degree $|i_X| = -1$). A straightforward corollary to a theorem of Fröhlicher–Nijenhuis (see [6]) states that, given a linear connection ∇ on M, the derivations of the form ∇_X, i_X generate the $\Omega(M)$-module $\mathrm{Der}\Omega(M)$.

The (super) differential 1-forms on $(M, \Omega(M))$ are defined as the duals $\mathrm{Der}^*\Omega(M)$, and k-forms are defined by taking exterior products as usual, and noting that they are *bigraded* objects; if, for instance, $\boldsymbol{\omega} \in \Omega^2(M, \Omega(M))$ (that is the way of denoting the space of 2-superforms), its action on two supervector fields $D, D' \in \mathrm{Der}\,\Omega(M)$ will be denoted $\left\langle D, D'; \boldsymbol{\omega}\right\rangle$, a notation well adapted to the fact that $\mathrm{Der}\,\Omega(M)$ is considered here as a left $\Omega(M)$-module and $\Omega^2(M, \Omega(M))$ as a right one. Other objects such as the graded exterior differential can be defined as in the classical setting, but taking into account the $\mathbb{Z}_2$-degree (for details in the spirit of this paper, see [16]). Thus, if $\alpha \in \Omega^0(M, \Omega(M))$, its graded differential $\mathbf{d}$ is given by $\langle D; \mathbf{d}\alpha\rangle = D(\alpha)$, and if $\boldsymbol{\beta} \in \Omega^1(M, \Omega(M))$, we have a 2-form $\mathbf{d}\boldsymbol{\beta} \in \Omega^2(M, \Omega(M))$ whose action is given by

$$\left\langle D, D'; \mathbf{d}\boldsymbol{\beta}\right\rangle = D(\left\langle D'; \boldsymbol{\beta}\right\rangle) - (-1)^{|D||D'|} D'(\langle D; \boldsymbol{\beta}\rangle) - \left\langle [D, D']; \boldsymbol{\beta}\right\rangle,$$

where $|D|$ denotes the degree of the derivation D.

[1]This is not a great loss of generality in view of the existence of the vector bundle isomorphism $TM \to E$, between TM and the Batchelor bundle, already mentioned in the Introduction (see [13]), so the changes needed to deal with the most general case are mainly notational.

3 Symplectic Supergeometry

A supersymplectic form is a non-degenerate[2] graded 2-form $\boldsymbol{\omega} \in \Omega^2(M, \Omega(M))$ such that $\mathbf{d}\boldsymbol{\omega} = 0$. Notice that there are two classes of supersymplectic forms: the even ones (for which $|\boldsymbol{\omega}|$ is even) act in such a way that, in terms of the induced $\mathbb{Z}_2$-degree,

$$|\langle D, D'; \boldsymbol{\omega}\rangle| = |D| + |D'|$$

and lead to symmetry properties similar to that of the non graded case, but the odd symplectic forms (for which $|\boldsymbol{\omega}|$ is odd) satisfy

$$|\langle D, D'; \boldsymbol{\omega}\rangle| = |D| + |D'| + 1.$$

As we will see below, these different properties translate into different symmetry properties of the symplectic Ricci tensors.

By the aforementioned result of Frölicher–Nijenhuis, given a linear connection ∇ on M, the study of the action of any 2-superform $\boldsymbol{\omega}$ can be reduced to that of a matrix of the type

$$\begin{pmatrix} \langle \nabla_X, \nabla_Y; \boldsymbol{\omega}\rangle & \langle \nabla_X, i_Y; \boldsymbol{\omega}\rangle \\ \langle i_X, \nabla_Y; \boldsymbol{\omega}\rangle & \langle i_X, i_Y; \boldsymbol{\omega}\rangle \end{pmatrix}$$

where $X, Y \in \mathscr{X}(M)$.

In the case of an odd symplectic form $\boldsymbol{\omega}$, this structure can be made more explicit as follows. Starting from a vector bundle isomorphim $H : TM \to T^*M$, we define an odd 1-form $\boldsymbol{\lambda}_H$, given by its action on basic derivations,

$$\begin{aligned} \langle \nabla_X; \boldsymbol{\lambda}_H\rangle &= H(X) \\ \langle i_X; \boldsymbol{\lambda}_H\rangle &= 0. \end{aligned}$$

(notice that this action is actually independent of ∇). Next, we define $\boldsymbol{\omega}_H$ by $\boldsymbol{\omega}_H = \mathbf{d}\boldsymbol{\lambda}_H$. Thus, the matrix of $\boldsymbol{\omega}_H$ now reads

$$\begin{aligned} \langle \nabla_X, \nabla_Y; \boldsymbol{\omega}_H\rangle &= (\nabla_X H)Y - (\nabla_Y H)X \\ \langle \nabla_X, i_Y; \boldsymbol{\omega}_H\rangle &= -H(X)(Y) \\ \langle i_X, \nabla_Y; \boldsymbol{\omega}_H\rangle &= H(Y)(X) \\ \langle i_X, i_Y; \boldsymbol{\omega}_H\rangle &= 0. \end{aligned} \tag{5}$$

In a sense, these are all the odd symplectic superforms, according to the following result.

Theorem 1 ([11]) *Let $\boldsymbol{\omega}$ be an odd symplectic form on $(M, \Omega(M))$, then there exist a superdiffeomorphism $\phi : \Omega(M) \to \Omega(M)$ and a fibre bundle isomorphism*

[2] In a technical sense that we will not describe here. See [12] for the details.

$H : TM \to T^*M$ *such that*

$$\phi^* \omega = \omega_H .$$

In what follows, we will restrict our attention to *odd* symplectic forms of the type ω_H. Let us insist that the reason is that even symplectic forms give rise to graded symmetric symplectic Ricci tensors (see [9] for details), and further contraction with the graded skew-symmetric symplectic form gives zero, thus leading to a trivial symplectic scalar supercurvature.

4 Fedosov Supermanifolds

Now that we know the essentials about the structure of supersymplectic forms, to begin the program sketched in Sect. 2 we need some facts about superconnections $\boldsymbol{\nabla}$ on $(M, \Omega(M))$. In particular, we will need the analog of the Levi-Cività theorem concerning the existence of superconnections such that $\boldsymbol{\nabla}\omega = 0$ for a supersymplectic form ω, and also their corresponding structure theorem. We follow here the approach in [14], although with some differences, the main one being that we do not assume that $\boldsymbol{\nabla}$ is adapted to the splitting H (also, see Theorem 3 below).

A superconnection on $(M, \Omega(M))$ is defined just as in the non-graded case, as an $\mathbb{R}$-bilinear mapping $\boldsymbol{\nabla} : \mathrm{Der}\, \Omega(M) \times \mathrm{Der}\, \Omega(M) \to \mathrm{Der}\, \Omega(M)$, whose action on (D, D') is denoted $\boldsymbol{\nabla}_D D'$, with the usual properties of $\Omega(M)$-linearity in the first argument and Leibniz's rule in the second[3]:

$$\boldsymbol{\nabla}_D(\alpha D') = D(\alpha)D' + (-1)^{|\alpha||D|}\alpha \boldsymbol{\nabla}_D D' .$$

The definition of torsion and curvature also mimics the non-graded case:

$$\langle D, D'; \mathrm{Tor}^{\boldsymbol{\nabla}} \rangle = \boldsymbol{\nabla}_D D' - (-1)^{|D||D'|}\boldsymbol{\nabla}_{D'} D - [D, D'],$$

and

$$\langle D, D', D''; \mathrm{Curv}^{\boldsymbol{\nabla}} \rangle = [\boldsymbol{\nabla}_D, \boldsymbol{\nabla}_{D'}]D'' - \boldsymbol{\nabla}_{[D,D']} D'',$$

where $[D, D'] = D \circ D' - (-1)^{|D||D'|} D' \circ D$, $[\boldsymbol{\nabla}_D, \boldsymbol{\nabla}_{D'}] = \boldsymbol{\nabla}_D \boldsymbol{\nabla}_{D'} - (-1)^{|D||D'|}$ $\boldsymbol{\nabla}_{D'} \boldsymbol{\nabla}_D$ are the graded commutators. As in the case of supersymplectic forms, we can describe a superconnection, once a linear connection ∇ on M is chosen, by a set of tensor fields characterizing its action on basic derivations,

$$\begin{aligned} \boldsymbol{\nabla}_{\nabla_X} \nabla_Y &= \nabla_{\nabla_X Y + K_0(X,Y)} + i_{L_0(X,Y)} \\ \boldsymbol{\nabla}_{\nabla_X} i_Y &= \nabla_{K_1(X,Y)} + i_{\nabla_X Y + L_1(X,Y)} \end{aligned}$$

[3] In particular, $\boldsymbol{\nabla}$ is not a tensor, hence the difference in notation.

$$\mathbf{\nabla}_{i_X}\nabla_Y = \nabla_{K_2(X,Y)} + i_{L_2(X,Y)}$$
$$\mathbf{\nabla}_{i_X}i_Y = \nabla_{K_3(X,Y)} + i_{L_3(X,Y)},$$

where $K_i, L_i : TM \otimes TM \to \Lambda T^*M \otimes TM$, for $i \in \{0, 1, 2, 3\}$. As a simplifying assumption, we will take a symmetric $\mathbf{\nabla}$. The relevant result is the following.

Theorem 2 ([14]) *Let ∇ be a linear connection on M. A superconnection $\mathbf{\nabla}$ on $(M, \Omega(M))$ is symmetric if and only if*

$$\begin{array}{ll} K_0(X,Y) = K_0(Y,X) - \mathrm{Tor}^{\nabla}(X,Y), & L_0(X,Y) = L_0(Y,X) + \mathrm{Curv}^{\nabla}(X,Y), \\ K_1(X,Y) = K_2(Y,X), & L_1(X,Y) = L_2(Y,X), \\ K_3(X,Y) = -K_3(Y,X), & L_3(X,Y) = -L_3(Y,X), \end{array} \tag{6}$$

for all $X, Y \in \mathscr{X}(M)$.

When the linear connection ∇ on M is symmetric, in the first equation of (6) we have,

$$K_0(X,Y) = K_0(Y,X),$$

and this will be assumed in the sequel.

The next step is to study those superconnections $\mathbf{\nabla}$ which are compatible with a given odd supersymplectic form $\boldsymbol{\omega}_H$, in the sense that $\mathbf{\nabla}\boldsymbol{\omega}_H = 0$. This amounts to saying that

$$D(\langle D_1, D_2; \boldsymbol{\omega}_H\rangle) = \langle \mathbf{\nabla}_D D_1, D_2; \boldsymbol{\omega}_H\rangle + (-1)^{|D||D_1|}\langle D_1, \mathbf{\nabla}_D D_2; \boldsymbol{\omega}_H\rangle,$$

for all $D, D_1, D_2 \in \mathrm{Der}\,\Omega(M)$. As a further simplifying assumption, we will take the linear connection ∇ compatible with the isomorphism $H : TM \to T^*M$, that is, $\nabla H = 0$ (so, (5) also gets modified). Then, we get the following result (which corrects the one appearing in [14]).

Theorem 3 ([9]) *A symmetric superconnection, $\mathbf{\nabla}$, is compatible with the odd symplectic form $\boldsymbol{\omega}_H$ if and only if*

(a) $H(K_3(X,Y), Z) = -H(K_3(X,Z), Y)$
(b) $H(K_2(X,Y), Z) = -H(Y, L_3(X,Z))$
(c) $H(X, L_2(Y,Z)) = H(Z, L_2(Y,X))$
(d) $H(K_1(X,Y), Z) = H(K_1(X,Z), Y)$
(e) $H(K_0(X,Y), Z) = -H(Y, L_1(X,Z))$
(f) $H(X, L_0(Y,Z)) = H(Z, L_0(Y,X))$,

for all $X, Y, Z \in \mathscr{X}(M)$.

It is a straightforward generalization of the corresponding result in the non-graded setting, that superconnections compatible with a given supersymplectic form exist and, moreover, they possess an affine structure (see [9] and, for a different approach [3]). Also generalizing the non-graded case [7], a Fedosov supermanifold is defined as a supermanifold endowed with a supersymplectic form and a compatible symmetric superconnection, see [8]. Combining Theorem 3 and (6) with (5), we get the following. Let ω_H be an odd supersymplectic form on $(M, \Omega(M))$, with $H : TM \to T^*M$ the associated bundle isomorphism. Let ∇ be a compatible, symmetric, linear connection on M (that is, $\nabla H = 0$), so the action of ω_H on basic derivations reads

$$\begin{aligned}\langle \nabla_X, i_Y; \omega_H \rangle &= -H(X)(Y)\\ \langle i_X, \nabla_Y; \omega_H \rangle &= H(Y)(X)\\ \langle \nabla_X, \nabla_Y; \omega_H \rangle &= 0 = \langle i_X, i_Y; \omega_H \rangle\,.\end{aligned} \tag{7}$$

Finally, let $\boldsymbol{\nabla}$ be a superconnection on $(M, \Omega(M))$, symmetric and compatible with ω_H, characterized by the tensors $K_i, L_i, i \in \{0, 1, 2, 3\}$. From the above results, a pair $((M, \Omega(M)), \boldsymbol{\nabla}, \omega_H)$ is a Fedosov supermanifold if and only if:

(g) K_0 is symmetric, L_0 satisfies $L_0(X, Y) = L_0(Y, X) + \mathrm{Curv}^{\nabla}(X, Y)$, and K_3, L_3 are skew-symmetric (from (6)).
(h) $K_1(X, Y) = K_2(Y, X)$ and $L_1(X, Y) = L_2(Y, X)$ (also from (6)).
(i) The above items (a) to (f) hold.

These conditions turn out to be very restrictive. From (b), (h) and (d), we get

$$-H(X, L_3(Y, Z)) = H(K_2(Y, X), Z) = H(K_1(X, Y), Z) = H(K_1(X, Z), Y)\,,$$

and, because of the skew-symmetry of L_3 (g), this equals

$$H(X, L_3(Z, Y)) = -H(K_2(Z, X), Y) = -H(K_1(X, Z), Y).$$

Thus, $H(K_1(X, Z), Y) = -H(K_1(X, Z), Y)$, which, in view of the fact that H is an isomorphism, leads to

$$K_1 = 0 = K_2$$

and, a posteriori,

$$L_3 = 0.$$

An immediate consequence is the following.

Corollary 1 *A symmetric superconnection $\boldsymbol{\nabla}$, compatible with the odd symplectic form ω_H, acts as*

$$\begin{aligned}\boldsymbol{\nabla}_{\nabla_X} \nabla_Y &= \nabla_{\nabla_X Y + K_0(X,Y)} + i_{L_0(X,Y)}\\ \boldsymbol{\nabla}_{\nabla_X} i_Y &= i_{\nabla_X Y + L_1(X,Y)}\\ \boldsymbol{\nabla}_{i_X} \nabla_Y &= i_{L_1(Y,X)}\\ \boldsymbol{\nabla}_{i_X} i_Y &= \nabla_{K_3(X,Y)},\end{aligned}$$

for any $X, Y \in \mathscr{X}(M)$.

Notice that such a $\mathbf{\nabla}$ is determined just by four ordinary tensor fields K_0, K_3, L_0, and L_1.

5 Odd Symplectic Scalar Curvature

To study the simplest case, we will start with an n-dimensional manifold M, an isomorphism $H : TM \to T^*M$ and a linear connection on M, ∇, such that $\nabla H = 0$. We also consider the odd symplectic form $\boldsymbol{\omega}$ (actually $\boldsymbol{\omega}_H$, but we suppress subindices for simplicity) given by (7) (denoting $H(X, Y) = H(X)(Y)$) and a compatible superconnection $\mathbf{\nabla}$ as in Corollary 1. Due to the symmetry properties of $\mathrm{Curv}^{\mathbb{W}}$, to characterize the action of the symplectic curvature tensor

$$\langle D_1, D_2, D_3, D_4; \mathbf{R}^{\omega}\rangle := \langle\, \langle D_1, D_2, D_3; \mathrm{Curv}^{\mathbb{W}}\rangle\,,\ D_4\ ;\ \boldsymbol{\omega}\rangle$$

it suffices to study the following cases, which define corresponding 7 tensor fields $A_1, \dots, A_5$, B_1, and B_3 (any other case gives a vanishing curvature) :

$$\begin{aligned}
\langle \nabla_X, \nabla_Y, \nabla_Z, \nabla_T\ ;\ \mathbf{R}^{\omega}\rangle &= H(T, B_1(X, Y, Z)) \\
\langle \nabla_X, \nabla_Y, \nabla_Z, i_T\ ;\ \mathbf{R}^{\omega}\rangle &= -H(A_1(X, Y, Z), T) \\
&= \langle \nabla_X, \nabla_Y, i_T, \nabla_Z\ ;\ \mathbf{R}^{\omega}\rangle \\
\langle \nabla_X, i_Y, \nabla_Z, \nabla_T\ ;\ \mathbf{R}^{\omega}\rangle &= H(T, B_3(X, Y, Z)) \\
&= -\langle i_Y, \nabla_X, \nabla_Z, \nabla_T\ ;\ \mathbf{R}^{\omega}\rangle \\
\langle \nabla_X, \nabla_Y, i_Z, i_T; \mathbf{R}^{\omega}\rangle &= -H(A_2(X, Y, Z), T) \\
\langle \nabla_X, i_Y, \nabla_Z, i_T; \mathbf{R}^{\omega}\rangle &= -H(A_3(X, Y, Z), T) \\
&= \langle \nabla_X, i_Y, i_T, \nabla_Z; \mathbf{R}^{\omega}\rangle \\
&= -\langle i_Y, \nabla_X, \nabla_Z, i_T; \mathbf{R}^{\omega}\rangle \\
&= -\langle i_Y, \nabla_X, i_T, \nabla_Z; \mathbf{R}^{\omega}\rangle \\
\langle \nabla_X, i_Y, i_Z, i_T; \mathbf{R}^{\omega}\rangle &= -H(A_4(X, Y, Z), T) \\
&= -\langle i_Y, \nabla_X, i_Z, i_T; \mathbf{R}^{\omega}\rangle \\
\langle i_X, i_Y, \nabla_Z, i_T; \mathbf{R}^{\omega}\rangle &= -H(A_5(X, Y, Z), T) \\
&= \langle i_X, i_Y, i_T, \nabla_Z; \mathbf{R}^{\omega}\rangle.
\end{aligned}$$

Of course, these new tensors can be explicitly computed from the K_i, L_i's. For instance, $A_2, A_3 \in \Gamma(T^*M \otimes T^*M \otimes T^*M \otimes T^*M \otimes TM)$, are given by

$$A_2(X, Y, Z)\cdot = -K_3(\mathrm{Curv}^{\nabla}(X, Y)\cdot, Z) \tag{8}$$

$$A_3(X, Y, Z)\cdot = -K_3(Y, L_0(X, Z)\cdot). \tag{9}$$

From these expression and items (a)–(i) above, we get the following [9].

Proposition 1 *If* $((M, \Omega(M)), \mathbf{\nabla}, \boldsymbol{\omega})$ *has the structure of a Fedosov supermanifold, then*

1. $A_3(X, Y, Z) = A_3(Z, Y, X) - A_2(X, Z, Y)$.

2. $H(A_3(X,Y,Z),T) = H(A_3(Z,Y,X),T) - H(A_2(Z,X,T),Y)$.
3. $H(A_3(Y,Z,X),T) = -H(A_3(Y,T,X),Z)$,

for any $X, Y, Z, T \in \mathscr{X}(M)$.

If some additional symmetry properties of H are added to these conditions, we get those symmetries of the Ricci tensor mentioned in the introduction, leading to a trivial scalar curvature as we will see below.

Corollary 2 *If H comes from a Riemannian metric or a symplectic form on M, then the graded Ricci tensor satisfies*

$$\left\langle \nabla_X, i_Y; \mathbf{Ric}^{\omega} \right\rangle = -\left\langle i_Y, \nabla_X; \mathbf{Ric}^{\omega} \right\rangle.$$

Finally, we proceed to compute the symplectic scalar curvature from a graded Ricci tensor with this property. To this end, we take a basis of homogeneous derivations $\{\nabla_{X_i}, i_{X_i}\}$ (where $\{X_i\}$, for $i \in \{1, \ldots, n\}$ is a local basis of vector fields on M). The odd supermatrix locally representing $\boldsymbol{\omega}$ has the form

$$\boldsymbol{\omega} = \begin{pmatrix} 0 & -H(X_i, X_j) \\ H(X_j, X_i) & 0 \end{pmatrix} = \begin{pmatrix} 0 & -H_{ij} \\ H^t_{ij} & 0 \end{pmatrix}.$$

Thus, the graded morphism induced by $\boldsymbol{\omega}$, $\boldsymbol{\omega}^\flat : \operatorname{Der} \Omega(M) \to \Omega^1(M, \Omega(M))$, has a supermatrix representative

$$\boldsymbol{\omega}^\flat = \begin{pmatrix} 0 & H_{ij} \\ -H^t_{ij} & 0 \end{pmatrix}.$$

This supermatrix is invertible, and its superinverse is readily found to be

$$(\boldsymbol{\omega}^\flat)^{-1} = \begin{pmatrix} 0 & -(H^t_{ij})^{-1} \\ (H_{ij})^{-1} & 0 \end{pmatrix}.$$

Now, the supermatrix associated to $\mathbf{Ric}^{\omega}$ has the structure

$$\mathbf{Ric}^{\omega} = \begin{pmatrix} A & B \\ C & D \end{pmatrix},$$

so

$$(\mathbf{Ric}^{\omega})^\flat = \begin{pmatrix} A^t & -(-1)^0 C^t \\ B^t & (-1)^0 D^t \end{pmatrix} = \begin{pmatrix} A^t & -C^t \\ B^t & D^t \end{pmatrix}.$$

The scalar curvature is defined by the supertrace of $\mathbf{Ric}^{\omega}$ with respect to $\boldsymbol{\omega}$; therefore, a straightforward computation shows that

$$\begin{aligned}\mathbf{Scal}^{\omega} &= \mathrm{STr}\left(\left(\omega^{\flat}\right)^{-1} \circ \left(\mathbf{Ric}^{\omega}\right)^{\flat}\right) \\ &= -\mathrm{Tr}\left(C^{t}\,(H_{ij})^{-1}\right) + \mathrm{Tr}\left(-B^{t}\,(H^{t}_{ij})^{-1}\right).\end{aligned}$$

Now, if H has a definite symmetry, from Corollary 2 we get $C = -B^{t}$ and consequently

$$\mathbf{Scal}^{\omega} = -\mathrm{Tr}\left(C^{t}\,(H_{ij})^{-1}\right) + \mathrm{Tr}\left(C\,(H^{t}_{ij})^{-1}\right).$$

But for any homogeneous invertible block A we have

$$(A^{t})^{-1} = (-1)^{|A|}(A^{-1})^{t}$$

(because, for homogeneous blocks, $(AB)^{t} = (-1)^{|A||B|}B^{t}A^{t}$), and also, because of the invariance of the trace under transpositions, $\mathrm{Tr}(A^{t}\,B) = \mathrm{Tr}(A\,B^{t})$, so

$$\mathbf{Scal}^{\omega} = -\mathrm{Tr}(C\,(H^{-1}_{ij})^{t}) + \mathrm{Tr}(C\,(H^{-1}_{ij})^{t}) = 0.$$

Thus, we deduce the following obstruction result (where we put back the subindex H for clarity).

Theorem 4 *If (M, H) is either a Riemannian or a symplectic manifold, then* $\mathbf{Scal}^{\omega_H} = 0$ *on* $(M, \Omega(M))$.

We believe that the preceding computations shed some light on the origin of the difficulties related to the construction of explicit examples of odd scalar supercurvatures (letting aside the question of their geometric meaning).

Let us finish by mentioning two possible ways of avoiding this obstruction. Of course, one consists in taking a general $H : TM \to T^{*}M$, not symmetric nor skew-symmetric. The problem here is that such objects are not as natural from the point of view of Physics as a metric or a symplectic form, and its introduction should be carefully justified. The other possibility involves the choice of a connection ∇ such that $\nabla H \neq 0$. This one is more interesting, as physically the choice of a connection is often part of the problem (for instance, in the Lagrangian version of Ashtekar's Canonical Gravity, connections are precisely the variables [10]). However, the study of this case is much more difficult and will be treated somewhere else [9].

Acknowledgments JM was supported by a Spanish Ministry of Science and Innovation project code MTM2012-33073. JV was supported by a CONACyT (México) project code CB-2012-179115.

References

1. Batalin, I.A., Bering, K.: Odd scalar curvature in field-antifield formalism. J. Math. Phys. **49**, 033515 (2008)
2. Batalin, I.A., Bering, K.: Odd scalar curvature in anti-Poisson geometry. Phys. Lett. B **663**(1–2), 132–135 (2008)

3. Blaga, P.A.: Symplectic connections on supermanifolds: existence and non-uniqueness. Studia Univ. Babes-Bolyai Math. **58**(4), 477–483 (2013)
4. De Castro, A., Martín, I., Quevedo, L., Restuccia, A.: Noncommutative associative superproduct for general supersymplectic forms. J. High Energy Phys. **0808**, 009 (2008)
5. Fedosov, B.V.: A simple geometrical construction of deformation quantization. J. Diff. Geom. **40**(2), 213–238 (1997)
6. Frlicher, A., Nijenhuis, A.: Theory of vector valued differential forms. Part I. Indag. Math. **18**, 338–360 (1956)
7. Gelfand, I., Retakh, V., Shubin, M.: Fedosov manifolds. Adv. Math. **136**(1), 104–140 (1998)
8. Geyer, B., Lavrov, P.M.: Fedosov supermanifolds: basic properties and the difference in even and odd cases. Int. J. Modern Phys. A **19**, 3195 (2004)
9. Hernández-Amador, R.: Ph.D. Thesis, Universidad de Sonora, México, in preparation (2015)
10. Jacobson, T., Smolin, L.: The left-handed spin connection as a variable for canonical gravity. Phys. Lett. B **196**, 39–42 (1987)
11. Kosmann-Schwarzbach, Y., Monterde, J.: Divergence operators and odd Poisson brackets. Ann. Inst. Fourier **52**, 419–456 (2002)
12. Kostant, B.: Graded manifolds, graded Lie theory, and prequantization. Differential Geometrical Methods in Mathematical Physics. Lecture Notes in Mathematics, vol. 570, pp. 177–306 (1977)
13. Monterde, J.: A characterization of graded symplectic structures. Diff. Geom. Appl. **2**, 81–97 (1992)
14. Monterde, J., Muñoz-Masqué, J., Vallejo, J.A.: The structure of Fedosov supermanifolds. J. Geom. Phys. **59**, 540–553 (2009)
15. Salgado, G., Vallejo, J.A.: The meaning of time and covariant superderivatives in supermechanics. Adv. Math. Phys. (2009) Article ID 987524
16. Vallejo, J.A.: Symplectic connections and Fedosov's quantization on supermanifolds. J. Phys. Conf. Ser. **343**, 012124 (2012)

Prime Submodules and Symmetric Algebras

Agustín Marcelo, Félix Marcelo and César Rodríguez

The commutative algebra investigation group of the University of Las Palmas de Gran Canaria is enormously grateful to Jaime Muñoz Masqué for being our advisor and true friend for all past years. Above all, Jaime deserves special mention for being a nice person always willing to share his expertise in several fields of mathematics. Thanks Jaime for your extraordinary generosity and talent in writing and teaching mathematics. We are pleased to take this opportunity to express our sincere thanks.

Abstract Prime submodules of a module N and its symmetric algebra $S(N)$ are used to study radicals of submodules and minimal components of symmetric algebras by translating results from one categorie to the another one and vice versa.

Keywords Prime submodules · Radical of submodules · Symmetric algebra · Minimal components

1 Introduction

Let R be a (commutative and unitary) ring and let N be an R-module. The intersection of all prime submodules of N containing a submodule $M \subset N$ is called the radical of M and it is denoted by $\mathrm{rad}_N(M)$ (see [8]). As is well-known, the radical $\sqrt{I}$ of an ideal $I \subset R$ is characterized as the set of elements $a \in R$ such that $a^n \in I$ for some $n \in \mathbb{Z}^+$. This result has stimulated several efforts to obtain a somewhat similar characterization for the radical of a submodule (see [2, 3, 8, 9]). In this article we

A. Marcelo · F. Marcelo · C. Rodríguez (✉)
Departmento de Matemáticas, Universidad de Las Palmas de Gran Canaria,
Campus de Tafira, 35017 Las Palmas de Gran Canaria, Spain
e-mail: cesar@dma.ulpgc.es

A. Marcelo
e-mail: amarcelo@dma.ulpgc.es

F. Marcelo
e-mail: fjmarcelo@gmail.com

M. Castrillón López et al. (eds.), *Geometry, Algebra and Applications: From Mechanics to Cryptography*, Springer Proceedings
in Mathematics & Statistics 161, DOI 10.1007/978-3-319-32085-4_13

first associate a prime ideal of the symmetric algebra of N—called the expansion of M—to each prime submodule $M \subset N$ and then we use it in order to obtain a characterization of the radical of a submodule, which is as follows: an element of a finitely generated R-module N belongs to $\mathrm{rad}_N(M)$ if and only if it is contained in the radical of the ideal of the symmetric algebra of N generated by all elements of M. As this result reduces the calculation of $\mathrm{rad}_N(M)$ to that of the radical of an ideal in a symmetric algebra, we apply our characterization to design an algorithm for computing some radicals of submodules of free modules by using the computer software package *CoCoA 3* (See [5, Sect. 3.2]).

On the other hand, by using as main tool the theory of prime submodules we describe the structure of the minimal prime ideals of the equidimensional symmetric algebra of a finitely generated module. We first characterize when the ideal generated by a submodule of a free module in the symmetric algebra is equidimensional and determine the minimal components of equidimensional symmetric algebras. Probably the most outstanding result about this question was obtained by Huneke and Rossi in [1, Sect. 3]. These authors showed, among other results, that if N is a finitely-generated module over a commutative Noetherian ring, then the symmetric algebra of N, $S(N)$, can have arbitrarily large number of minimal components and they tried to identify the prime ideals of R which are the contraction of a minimal prime of $S(N)$.

Let R be a universally catenary Noetherian domain and let N be a finitely generated R-module such that $S(N)$ is equidimensional. If $\mathfrak{p}$ is a prime ideal of R, then the least number of generators of $N_\mathfrak{p}$ is denoted by $\nu(N_\mathfrak{p})$. Let $f\colon SpecS(N) \to Spec(R)$ denote the induced natural map. Then, given $\mathfrak{p} \in Spec(R)$, our basic purpose is to prove that in $f^{-1}(\mathfrak{p})$ there is a minimal prime ideal of $S(N)$ if and only if $\nu(N_\mathfrak{p}) - ht\mathfrak{p} = rankN$. If $\mathfrak{p}$ fulfils this condition, then there exists a unique minimal prime ideal of $S(N)$ in $f^{-1}(\mathfrak{p})$, denoted by $\mathscr{E}_{\mathfrak{p}(0)}$ and defined as

$$\mathscr{E}_{\mathfrak{p}(0)} = \{b \in S(N) : ab \in \mathfrak{p} \cdot S(N) \text{ for some } a \in R - \mathfrak{p}\}.$$

2 Preliminaries

Let R be a commutative Noetherian ring with identity and let N be a finitely generated R-module. If one tries to generalize the concept of a prime ideal (resp. primary) from R to N, one is led to the following

Definition 1 Recall that a proper submodule P of N is said to be a prime (resp. primary) submodule if for every $a \in R$, the induced homothety $N/P \xrightarrow{\cdot a} N/P$ is either injective or null (resp. nilpotent).

In light of this definition, it turns out that if P is a prime (resp. primary) submodule of N then the set of homotheties of R vanishing on N/P, i.e.,

$$(P : N) = \{a \in R \;/\; aN \subseteq P\} = Ann(N/P)$$

is a prime (resp. primary) ideal of R. Furthermore, if P is a primary submodule of N, the radical of the primary ideal $(P : N)$, denoted by $\sqrt{(P : N)}$, is a prime ideal of R formed by all nilpotent homotheties of R on N/P, i.e., $\sqrt{(P : N)} = \{a \in R \mid a^n N \subseteq P \text{ for some } n > 0\}$. Thus if P is a prime submodule of N with $\mathfrak{p} = (P : N)$ we shall call P a $\mathfrak{p}$-prime submodule and if P is a primary submodule of N being $\mathfrak{p} = \sqrt{(P : N)}$ we will say that P is a $\mathfrak{p}$-primary submodule. Note that a $\mathfrak{p}$-primary submodule P of N is $\mathfrak{p}$-prime if and only if $(P : N) = \mathfrak{p} \in SpecR$.

Definition 2 Let L be a proper submodule of a R-module N. Given a prime ideal $\mathfrak{p}$ of R, we will denote by $\mathfrak{p}(L)$ the following submodule of N:

$$\mathfrak{p}(L) = \{n \in N : an \in L + \mathfrak{p}N, \text{ for some } a \in R - \mathfrak{p}\}.$$

With the above notations, it is easy to see that either $\mathfrak{p}(L) = N$ or $\mathfrak{p}(L)$ is a $\mathfrak{p}$-prime submodule of N, which is contained in every $\mathfrak{p}$-prime submodule of N containing L.

Definition 3 Let R be a Noetherian domain. Let N be a finite R-module. A submodule M of N is said to be a 0-prime submodule if N/M is a torsion-free R-module or, equivalently, if zero is the unique noninjective homothety on N/M.

Definition 4 Let R be a (commutative and unitary) ring and let N be an R-module. The intersection of all prime submodules of N containing a submodule $M \subset N$ is called the radical of M and it is denoted by $\text{rad}_N(M)$ (see [8]).

Let $S(N) = \oplus_{i \geq 0} S^i(N)$ be the symmetric algebra of an R-module N endowed with its natural $\mathbb{Z}$-graduation. Throughout this paper we identify $S^0(N)$ (resp. $S^1(N)$) to R (resp. N).

Definition 5 *Let N be a finitely generated R-module and let M be a prime submodule of N. We define the* expansion $\mathscr{E}_M$ *of M to be the set of all elements $b \in S(N)$ for which there exists $a \in R$, $a \notin \mathfrak{p}_M$ such that $a \cdot b \in (\mathfrak{p}_M, M) \cdot S(N)$.*

Proposition 1 *With the assumptions and notations above we have*

1. $\mathscr{E}_M$ is a prime ideal of $S(N)$.
2. $\mathscr{E}_M \cap R = \mathfrak{p}_M$, $\mathscr{E}_M \cap N = M$.
3. $\mathscr{E}_M$ is a homogeneous ideal; *i.e.*, $\mathscr{E}_M = \oplus_{i \geq 0} \mathscr{E}^i_M$, $\mathscr{E}^i_M = \mathscr{E}_M \cap S^i(N)$.
4. The mapping $M \mapsto \mathscr{E}_M$ is an injection from the set of prime submodules of N into $\text{Spec}S(N)$. (See [4, Sect. 2])

3 Computing the Radical of a Submodule

In [9] it is shown that if R is a principal ideal domain and N a finitely generated R-module then the radical of every submodule $M \subseteq N$ coincides with the submodule generated by its envelope; that is, $\text{rad}_N(M) = \langle E(M) \rangle$, where $E(M)$ is the set of all $x \in N$ for which there exist $a \in R$, $y \in N$ such that $x = a \cdot y$ and $a^n y \in M$ for some $n \in \mathbb{Z}^+$. In this case the module N is said to satisfy the radical formula (in short, N s.t.r.f). In [2] this result has been extended to any Dedekind domain R and any R-module.

Now we obtain the following characterization of the M-radical of a submodule which ensures that the radical of a submodule coincides to the radical of an ideal of the symmetric algebra.

Theorem 1 *Let R be a ring, let N be a finitely generated R-module and let $Q \subseteq N$ be a submodule. An element $x \in N$ belongs to $\mathrm{rad}_N(Q)$ if and only if $x \in \sqrt{Q \cdot S(N)}$, or equivalently, $x^n \in Q \cdot S(N)$ for some $n \in \mathbb{Z}^+$.*

Proof First, assume $x \in \mathrm{rad}_N(Q)$. Let $\mathfrak{p} \in \mathrm{Spec} S(N)$ such that $Q \cdot S(N) \subseteq \mathfrak{p}$. If we set $M = \mathfrak{p} \cap N$, by applying [4, Proposition 1.2], it is easy to see that M is a prime submodule of N. Moreover, since $Q \subseteq M$ we have $x \in M$. Hence $x \in \mathfrak{p}$ and consequently $x \in \sqrt{Q \cdot S(N)}$.

Conversely, let $x \in N$ be an element such that $x^n \in Q \cdot S(N)$ for some $n \in \mathbb{Z}^+$. We must show that if M is a prime submodule of N such that $Q \subseteq M$, then $x \in M$. Indeed, let $\mathscr{E}_M$ be the expansion of M. Since $Q \subseteq M$ we have $Q \cdot S(N) \subseteq \mathscr{E}_M$ so that $x^n \in \mathscr{E}_M$ and since $\mathscr{E}_M$ is a prime ideal we obtain $x \in \mathscr{E}_M$. Hence $x \in \mathscr{E}_M \cap N = M$ and the result is proved.

Next we are going to apply the above result to calculate some radicals of submodules.

From now on, we denote by $A = R[x_1, \ldots, x_n] = \bigoplus_{i \geq 0} A(i)$ the positively graded ring of all polynomials over a ring R. If N is an R-module, then there is an exact sequence of R-modules

$$0 \longrightarrow K \longrightarrow F \stackrel{\pi}{\longrightarrow} N \longrightarrow 0 \tag{1}$$

where F is a free R-module. Given a proper submodule $M \subset N$ we set $L = \pi^{-1}(M)$. If $\{e_1, \ldots, e_n\}$ is a basis of F, we have an isomorphism $\varphi \colon F \longrightarrow A(1)$, $\varphi(e_i) = x_i$, $1 \leq i \leq n$. We denote by I, J the ideals $\varphi(K) \cdot A$, $\varphi(L) \cdot A$, respectively. Clearly $I \subseteq J$ and from (1) we obtain an exact sequence

$$0 \longrightarrow I \longrightarrow A \longrightarrow S(N) \longrightarrow 0.$$

Moreover, I, J and $\sqrt{J}$ are homogeneous ideals with gradings $I = \oplus_{k \geq 0} I(k)$, $J = \oplus_{k \geq 0} J(k)$, and $\sqrt{J} = \oplus_{k \geq 0} \sqrt{J}(k)$, respectively.

Proposition 2 *With the above notations, the following equality holds true*

$$\mathrm{rad}_N(M) = \sqrt{J}(1)/I(1).$$

Proof From the exact sequence (1) we have $\mathrm{rad}_N(M) = \mathrm{rad}_F(L)/K$. Moreover, since $\varphi(K) = I(1)$ and $\varphi(L) = J(1)$, we can apply Theorem 1 to obtain $\mathrm{rad}_N(M) = \sqrt{J}(1)/I(1)$.

By using the above result we can calculate some radicals of submodules of free modules.

Let $R = k[y_1, \dots, y_n]$ be the ring of polynomials over a field k, let $A = R[x_1, \dots, x_n]$ and let M be a submodule of a free module F generated by $\{e_1, \dots, e_n\}$. By using the isomorphism $\varphi\colon F \longrightarrow A(1)$ above it turns out that $J = \varphi(M) \cdot A$. So if we know the generators of the submodule M, we only need to replace e_i by x_i to obtain the generators of the ideal J. Once we know this, by using computer algebra systems it is possible to obtain the generators of the ideal $\sqrt{J}$. Finally, taking into account the preceding Proposition it follows that the $\mathrm{rad}_F(M)$ is spanned by the linear generators of $\sqrt{J}$ in the variables $x_1, \dots, x_n$. To illustrate the whole process we present the following

Example 1 Let $R = \mathbb{Q}[x, y, z]$ be the polynomial ring over $\mathbb{Q}$ in three variables and let F be a free R-module with basis $\{e_1, \dots, e_5\}$. Let M be the following submodule of F:

$$M = \left\langle x^2e_1 + y^2e_2,\ x^2ze_2 + y^3e_3,\ y^3ze_3 + x^4e_4,\ xz^3e_4 + y^4e_5 \right\rangle.$$

By replacing $\{e_1, \dots, e_5\}$ by $\{a, b, c, d, e\}$ respectively, we shall consider the ideal

$$I = \left\langle x^2a + y^2b,\ x^2zb + y^3c,\ y^3zc + x^4d,\ xz^3d + y^4e \right\rangle.$$

Next, according to the computation made by T. Recio using the package Radical, by M. Caboara, implemented in CoCoA 3.5, we obtain that the $\sqrt{I}$ is given by

$$\begin{aligned}\sqrt{I} = \langle & x^2a + y^2b,\ ycd^2 + xabe,\ xcd^2 - yb^2e,\ xzd^2 - y^2ae,\ -yzac + x^2bd,\\ & yzb^2 - y^2ac,\ xzb^2 - xyac,\ yz^2b - x^2yd,\ xz^2b - x^3d,\ x^2zb + y^3c,\\ & xz^2a + xy^2d,\ xyd^3 + yza^2e,\ yz^2ae + y^3de,\ x^2d^3 + xza^2e,\\ & yzbd^2 + xya^2e,\ yz^2cd - xy^2be,\ xz^2cd - x^2ybe, xz^3d + y^4e,\\ & xza^2c + xyb^2d,\ y^3zc + x^4d,\ xb^2d^3 + ya^3ce,\ xa^3c^2 + xb^4d,\\ & ya^3c^2 + yb^4d,\ yz^4c - x^3y^2e,\ ybd^5 - ya^4e^2,\ xbd^5 - xa^4e^2,\\ & y^2bd^4 - xza^3de,\ yb^3d^3 - xa^4ce,\ -y^2a^2c^3d + xyb^5e\rangle.\end{aligned}$$

Thus the radical of the submodule M of F is

$$\begin{aligned}rad_F(M) = \langle & x^2e_1 + y^2e_2,\ yz^2e_2 - x^2ye_4,\ xz^2e_2 - x^3e_4,\ x^2ze_2 + y^3e_3,\\ & xz^2e_1 + xy^2e_4,\ xz^3e_4 + y^4e_5,\ y^3ze_3 + x^4e_4,\ yz^4e_3 - x^3y^2e_5\rangle\end{aligned}$$

4 Determining Minimal Components of Equidimensional Symmetric Algebras

This section is devoted to describing minimal components of equidimensional symmetric algebras but to do it we need the following preliminaries results (see [6], Sect. 2):

Proposition 3 *Let $(R, \mathfrak{p})$ be a Noetherian local ring, let F be a free R-module of rank n and let M be a proper submodule of F. Then either $\mathfrak{p}(M) = \mathfrak{p}F$ or $\mathfrak{p}(M) = \mathfrak{p}F + (e_1, \ldots, e_h)$, where $e_1, \ldots, e_h$ form a part of a basis of F.*

Remark 1 From now on, we assume that h is the largest possible value.

Proposition 4 *Let $(R, \mathfrak{p})$, F and M be as above. Then, either*

$$\mathscr{E}_{\mathfrak{p}(M)} = \mathfrak{p}S(F)$$

or

$$\mathscr{E}_{\mathfrak{p}(M)} = \mathfrak{p}S(F) + (x_1, x_2, \ldots, x_h)S(F),$$

where $S(F) = R\left[x_1, x_2, \ldots, x_h, x_{h+1}, \ldots, x_n\right]$.

Proposition 5 *Let $(R, \mathfrak{p})$, F, and M be as above. Then, either*

$$tr.\deg_{k(\mathfrak{p})} k(\mathscr{E}_{\mathfrak{p}(M)}) = n \;\; if \;\; \mathscr{E}_{\mathfrak{p}(M)} = \mathfrak{p}S(F),$$

or

$$tr.\deg_{k(\mathfrak{p})} k(\mathscr{E}_{\mathfrak{p}(M)}) = n - h \;\; if \;\; \mathscr{E}_{\mathfrak{p}(M)} = \mathfrak{p}S(F) + (x_1, \ldots, x_h)S(F).$$

Again, let $(R, \mathfrak{p})$, F and M be as above. Let us consider the short exact sequence $0 \longrightarrow M \longrightarrow F \xrightarrow{\pi} N \longrightarrow 0$ and let $\nu(N)$ be the minimal number of generators of N. As we have seen, if $M + \mathfrak{p}F = \mathfrak{p}F + (e_1, e_2, \ldots, e_h)$, then there exists a basis $\{e_1, e_2, \ldots, e_h, e_{h+1}, \ldots, e_n\}$ of F. Thus, by Nakayama's Lemma, $\nu(N) = n - h$.

Proposition 6 *With the same hypotheses and notations as above,*

$$tr.\deg_{k(\mathfrak{p})} k(\mathscr{E}_{\mathfrak{p}(M)}) = \nu(N).$$

4.1 *Equidimensional Ideals*

Let R be a Noetherian domain, let F be a free R-module of finite rank, and let $S(F)$ be the symmetric algebra of F. In what follows, we assume that every irredundant chain of prime ideals of $S(F)$ has the same length.

Remark 2 Let M be a submodule of F and denote by $M \cdot S(F)$ the ideal generated by M in $S(F)$. As is well known, the ideal $M \cdot S(F)$ is said to be equidimensional if all its minimal prime ideals have the same codimension. These ideals are interesting because the quotient $S(F)/M \cdot S(F)$ is an equidimensional symmetric algebra.

Proposition 7 *Every minimal prime ideal over* $M \cdot S(F)$ *is the expansion* $\mathscr{E}_{\mathfrak{p}(M)}$ *for some* $\mathfrak{p} \in Spec(R)$.

Proof Let $I \in Spec S(F)$ be a minimal prime ideal over $M \cdot S(F)$ and set $\mathfrak{p} = I \cap R$. Then it is easily shown that $P = I \cap F$ is a $\mathfrak{p}$-prime submodule containing M. Hence $\mathfrak{p}(M) \subseteq P$ and we have $\mathscr{E}_{\mathfrak{p}(M)} \subseteq \mathscr{E}_P \subseteq I$. Since I is a minimal ideal over $M \cdot S(N)$ and $M \cdot S(N) \subseteq \mathscr{E}_{\mathfrak{p}(M)}$ it is deduced that $\mathscr{E}_{\mathfrak{p}(M)} = I$ and the desired equality follows.

Proposition 8 *With the same notations as above, let* $M \subset F$ *be a submodule and let* $r_{\mathfrak{p}}(M)$ *be the greatest rank of a free direct summand of* $F_{\mathfrak{p}}$ *contained in* $M_{\mathfrak{p}}$, $\mathfrak{p} \in Spec(R)$. *Then,*

$$ht\mathscr{E}_{\mathfrak{p}(M)} = ht\mathfrak{p} + r_{\mathfrak{p}}(M).$$

Proof From the properties of localization it follows

$$ht\mathscr{E}_{\mathfrak{p}(M)} = \left(ht\mathscr{E}_{\mathfrak{p}(M)}\right)_{\mathfrak{p}}.$$

On the other hand, we have

$$M_{\mathfrak{p}} = M'_{\mathfrak{p}} \oplus l_{\mathfrak{p}}(M),$$

where $l_{\mathfrak{p}}(M)$ is a direct summand of $F_{\mathfrak{p}}$ contained in M_p of rank $r_{\mathfrak{p}}(M)$ and $M'_{\mathfrak{p}}$ is a submodule of $M_{\mathfrak{p}}$. This implies that $M'_{\mathfrak{p}} \subseteq \mathfrak{p}F_{\mathfrak{p}}$ since if an element $m' \in M'_{\mathfrak{p}}$ is not included in $\mathfrak{p}F_{\mathfrak{p}}$ then (m') would be a direct summand of $F_{\mathfrak{p}}$ as follows from Nakayama's lemma. Hence

$$M_{\mathfrak{p}} = M''_{\mathfrak{p}} \oplus (m') \oplus l_{\mathfrak{p}}(M),$$

thus contradicting the greatest rank of $l_{\mathfrak{p}}(M)$. Now it is not difficult to see that

$$\mathfrak{p}(M_{\mathfrak{p}}) = \mathfrak{p}F_p \oplus l_{\mathfrak{p}}(M).$$

By identifying, as usual, F to $S_1(F)$ it follows that $\mathfrak{p}F_p \oplus l_{\mathfrak{p}}(M)$ generates an ideal of $S(F)_{\mathfrak{p}}$ whose height is just $ht\mathfrak{p} + r_{\mathfrak{p}}(M)$. Again by properties of localization we obtain that the precedent height coincides with the height of the ideal $\mathscr{E}_{\mathfrak{p}(M)}$, and so the proof is completed.

Before passing to the statement of our next result, we need to prove the following lemma:

Lemma 1 $\mathscr{E}_{0(M)}$ *is a minimal prime ideal over* $M \cdot S(M)$ *such that*

$$ht\mathscr{E}_{0(M)} = rank(M).$$

Proof First assume that $\mathscr{E}_{0(M)}$ is not minimal over $M \cdot S(M)$. Let I be a prime ideal of $S(F)$ such that

$$M \cdot S(F) \subseteq I \underset{\neq}{\subset} \mathscr{E}_{0(M)}.$$

If $I \cap R = (0)$ it is not difficult to see that $I = \mathscr{E}_{0(M)}$, contrary to the initial assumption. Thus let $I \cap R = \mathfrak{p} \neq (0)$. Since $\mathscr{E}_{0(M)} \cap R = (0)$, after localizing $S(F)$ by the multiplicative set $S = R - (0)$ it turns out that $(\mathscr{E}_{0(M)})_{(0)}$ is a proper ideal of $S(F)_{(0)}$ and the same happens for the ideal $I_{(0)}$ since $I \underset{\neq}{\subset} \mathscr{E}_{0(M)}$. But in $S(F)_{(0)}$ we have $\mathfrak{p}_{(0)} = R_{(0)}$ it follows that $I_{(0)}$ contains the identity element. Therefore $I_{(0)} = S(F)_{(0)}$ which leads us to a contradiction. Hence $I = \mathscr{E}_{0(M)}$. On the other hand, $S(F)_{(0)}$ is a polynomial ring over the field $R_{(0)}$ in which $M \cdot S(F)_{(0)}$ is a prime ideal whose height is just $rank(M)$. Finally, since $M \cdot S(F)_{(0)} = (\mathscr{E}_{0(M)})_{(0)}$ we deduced the desired result taking into account that $ht\mathscr{E}_{0(M)} = ht(\mathscr{E}_{0(M)})_{(0)}$.

Theorem 2 *With the same notations, the following conditions are equivalent:*

1. *$M \cdot S(F)$ is an equidimensional ideal.*
2. *If $\mathscr{E}_{\mathfrak{p}(M)}$ is a minimal prime ideal over $M \cdot S(F)$, then*

$$ht\mathfrak{p}+r_{\mathfrak{p}}(M) = rank(M).$$

Proof Assume $M \cdot S(F)$ is an equidimensional ideal and let I be a minimal prime ideal over $M \cdot S(M)$. By virtue of Proposition 7 the ideal I is the expansion of a prime submodule $\mathfrak{p}(M)$ where by definition we have

$$I \cap F = \mathfrak{p}(M) \text{ and } I = \mathscr{E}_{\mathfrak{p}(M)}.$$

By applying Proposition 8 it now follows that

$$htI = ht\mathfrak{p}+r_{\mathfrak{p}}(M).$$

On the other hand, by localizing $M \cdot S(F)$ in the generic point of R and by contracting this localization to $S(F)$ we obtain just the ideal $\mathscr{E}_{0(M)}$.

Using now the precedent lemma and the hypothesis of equidimensionality of $M \cdot S(F)$, we deduce that

$$ht\mathfrak{p}+r_{\mathfrak{p}}(M) = rank(M).$$

Let us assume now that for every minimal prime ideal $\mathscr{E}_{\mathfrak{p}(M)}$ over $M \cdot S(F)$ is $ht\mathfrak{p}+r_{\mathfrak{p}}(M) = rank(M)$. Clearly in this case $M \cdot S(F)$ is equidimensional and we can conclude.

4.2 Minimal Components of $S(N)$

We first need the following

Lemma 2 *Let R be a universally catenary Noetherian domain, let $\mathfrak{p}$ be a prime ideal of R, let N be a finitely generated R-module and let*

$$0 \longrightarrow M \cdot S(F) \longrightarrow S(F) \xrightarrow{\lambda} S(N) \longrightarrow 0$$

be the exact sequence induced from the exact sequence of R-modules,

$$0 \longrightarrow M \longrightarrow F \xrightarrow{\pi} N \longrightarrow 0,$$

in which F is free. Then, $\mathscr{E}_{\mathfrak{p}(0)}$ is a minimal prime ideal of $S(N)$ if and only if $\mathscr{E}_{\mathfrak{p}(M)}$ is a minimal prime ideal over $M \cdot S(F)$.

Proof Assume that $\mathscr{E}_{\mathfrak{p}(0)}$ is a minimal prime ideal of $S(N)$. In this case, we can easily see that $\lambda^{-1}(\mathscr{E}_{\mathfrak{p}(0)})$ is a minimal prime ideal over $M \cdot S(F)$ and $\lambda^{-1}\mathscr{E}_{\mathfrak{p}(0)} = \mathscr{E}_{\mathfrak{p}(M)}$.

Conversely, if $\mathscr{E}_{\mathfrak{p}(M)}$ is a minimal prime ideal of $S(F)$ over $M \cdot S(F)$, then $\lambda(\mathscr{E}_{\mathfrak{p}(M)}) = \mathscr{E}_{\mathfrak{p}(0)}$ is also a minimal prime in $S(N)$.

Theorem 3 *With the same notations and assumptions as in the previous lemma, $\mathscr{E}_{\mathfrak{p}(0)}$ is a minimal prime ideal of $S(N)$ if and only if $rank N = \nu(N_{\mathfrak{p}}) - ht\mathfrak{p}$.*

Proof Let $T = T(S(N))$ be the torsion R-module of $S(N)$. By [1, p. 201] T is a minimal prime ideal of $S(N)$. Therefore $\lambda^{-1}(T)$ is a prime ideal of $S(F)$ minimal over $M \cdot S(F)$. Moreover, we have $\lambda^{-1}(T) = \mathscr{E}_{0(M)}$, where $\mathscr{E}_{0(M)}$ is the expansion of the 0-prime submodule $0(M)$, i.e.,

$$\mathscr{E}_{0(M)} = \{b \in S(F) : ab \in MS(F) \text{ for some } a \neq 0\}.$$

Assume now $\mathscr{E}_{\mathfrak{p}(M)}$ is a minimal prime ideal over $M \cdot S(F)$. Taking into account that $S(F)$ is a catenary ring because R is universally catenary and by hypothesis $S(N)$ is equidimensional, it is not difficult to see that all minimal prime ideals over $M \cdot S(F)$ have the same height. Then $ht\mathscr{E}_{\mathfrak{p}(M)} = ht\mathscr{E}_{0(M)}$ (see [7, p. 118]). On the other hand, by applying [3, Theorem 15.5, p. 118] is obtained

$$ht\mathscr{E}_{0(M)} = tr.\deg_{k(0)} S(F)_{(0)} - tr.\deg_{k(0)} k(\mathscr{E}_{0(M)}),$$

where $k(0)$ is the field of fractions of R.

Next, from Proposition 4, we have

$$tr.\deg_{k(0)} k(\mathscr{E}_{0(M)}) = \nu(N_{(0)}) = rank N.$$

Hence

$$ht\mathscr{E}_{0(M)} = tr.\deg_{k(0)} S(F)_{(0)} - rankN.$$

Again by ([7, p. 118]) we have

$$ht\mathscr{E}_{\mathfrak{p}(M)} = tr.\deg_{k(0)} S(F)_{(0)} + ht\mathfrak{p} - tr.\deg_{k(\mathfrak{p})} k(\mathscr{E}_{\mathfrak{p}(M)}),$$

thus

$$ht\mathscr{E}_{\mathfrak{p}(M)} = tr.\deg_{k(0)} S(F)_{(0)} + ht\mathfrak{p} - \nu(N_{\mathfrak{p}}),$$

which implies that $rankN = \nu(N_{\mathfrak{p}}) - ht\mathfrak{p}$.

Conversely, suppose that $rankN = \nu(N_{\mathfrak{p}}) - ht\mathfrak{p}$. By applying [7, p. 118] we obtain

$$\begin{aligned} ht\mathscr{E}_{\mathfrak{p}(M)} &= ht\mathfrak{p}+tr.\deg_{k(0)} S(F)_{(0)} - tr.\deg_{k(\mathfrak{p})} k(\mathscr{E}_{0(M)}) \\ &= ht\mathfrak{p}+tr.\deg_{k(0)} S(F)_{(0)} - \nu(N_{\mathfrak{p}}), \end{aligned}$$

On the other hand,

$$\begin{aligned} ht\mathscr{E}_{0(M)} &= tr.\deg_{k(0)} S(F)_{(0)} - tr.\deg_{k(0)} k(\mathscr{E}_M) \\ &= tr.\deg_{k(0)} S(F)_{(0)} - \nu(N_{(0)}) \\ &= tr.\deg_{k(0)} S(F)_{(0)} - rankN. \end{aligned}$$

Then, by virtue of hypothesis $ht\mathscr{E}_{\mathfrak{p}(M)} = ht\mathscr{E}_{0(M)}$. By using the fact that $S(F)$ is a catenarian ring it is deduced that $\mathscr{E}_{\mathfrak{p}(M)}$ is a minimal prime ideal over $M \cdot S(F)$ in $S(F)$ since $\mathscr{E}_{\mathfrak{p}(0)}$ is minimal in $S(N)$.

References

1. Huneke, C., Rossi, M.: The dimension and components of symmetric algebras. J. Algebra **98**, 200–210 (1986)
2. Jenkins, J., Smith, P.F.: On the prime radical of a module over a commutative ring. Commun. Algebra **20**(12), 3593–3602 (1992)
3. Chin-Pi, Lu: M-radicals of submodules in modules. Math. Japonica **34**(2), 21–219 (1989)
4. Marcelo, A., Muñoz, J.: Masqué, prime submodules, the descent invariant, and modules of finite length. J. Algebra **189**, 273–293 (1997)
5. Marcelo, A., Rodríguez, C.: Radicals of submodules and symmetric algebra. Commun. Algebra **28**(10), 4611–4617 (2000)
6. Marcelo, A., Marcelo, F., Rodríguez, C.: Equidimensional symmetric algebras. J. Korean Math. Soc. **47**(2), 289–297 (2010)
7. Matsumura, H.: Commutative Ring Theory. Cambridge University Press, Cambridge (1986)
8. McCasland, R., Moore, M.: On radicals of submodules of finitely generated modules. Can. Math. Bull. **29**(1), 37–39 (1986)
9. McCasland, R., Moore, M.: On radicals of submodules. Commun. Algebra **19**(5), 1327–1341 (1991)

Application to Cybersecurity of the Stability Theory of the Systems of Ordinary Differential Equations

Ángel Martín del Rey and Gerardo Rodríguez Sánchez

One machine can do the work of fifty ordinary men. No machine can do the work of one extraordinary men (E. Hubbard).
Dedicated to Jaime Muñoz Masqué, our mentor and friend, on the occasion of his 65th birthday.

Abstract The main goal of this work is to show an application of the stability theory of systems of ordinary differential equations to cybersecurity. Specifically, we will focus our attention on the study of the systems used in the mathematical models to simulate malware spreading on computer networks. Thus, a compartmental SCIRS model for computer worms spreading is proposed and analyzed.

Keywords Malware propagation · Differential equations · Stability

1 Introduction

In the last three decades the scientific and technological progress of our society has been enormous, which is due to the development of the information and communication technologies.

It is safe to say that the great majority of our relationships depend on these technologies, so that the so-called e-society becomes a reality. This suggestive scenario, which managed in a timely manner can lead to high levels of welfare, it is not without risks and dangers. Consequently, it is very important to manage their security in an effective way. In this sense we can highlight the development of new and

Á. Martín del Rey (✉) · G. Rodríguez Sánchez
Department of Applied Mathematics, Institute of Fundamental Physics and Mathematics, University of Salamanca, Salamanca, Spain
e-mail: delrey@usal.es

G. Rodríguez Sánchez
e-mail: gerardo@usal.es

M. Castrillón López et al. (eds.), *Geometry, Algebra and Applications: From Mechanics to Cryptography*, Springer Proceedings in Mathematics & Statistics 161, DOI 10.1007/978-3-319-32085-4_14

increasingly sophisticated malware specimens whose economic and social effects can be very serious [2].

Cybersecurity is the branch of science that deals with protection (both reactively and proactively) of the information that are stored, managed and transmitted electronically via different computer networks. The design and development of security protocols involve different disciplines which include mathematical modeling. This plays a very important role in the study of malware spreading on computer networks. The great majority of mathematical models proposed to date to simulate this phenomenon are based on the use of systems of ordinary differential equations [3]. Its mathematical analysis allows us to draw conclusions that make us understand better the propagation mechanisms and design control strategies to minimize the malicious behavior of malware.

The main goal of this work is to show the use of the stability theory of (autonomous) systems of ordinary differential equations in analyzing the behavior of the dynamic of the last mentioned mathematical models. To achieve this goal, we propose a new model to simulate the spreading of a computer worm in a computer network such that the local stability of their equilibrium points is studied.

The rest of the paper is organized as follows: in Sect. 2 the basic notions on the stability theory of the systems of (three) ordinary differential equations are introduced; the detailed description of the proposed model and the study of its local stability is shown in Sect. 3; finally, in Sect. 4 the main conclusions are stated.

2 Mathematical Background

Set

$$\begin{cases} x'(t) = f(x, y, z) \\ y'(t) = g(x, y, z) \\ z'(t) = h(x, y, z) \end{cases} \quad (1)$$

an autonomous system of ordinary differential equations such that $f, g, h \in \mathscr{C}^1(\Gamma)$, where Γ is the feasible region. The point $e^* = (x^*, y^*, z^*) \in \Gamma$ is said to be an *equilibrium point* of (1) if:

$$f\left(x^*, y^*, z^*\right) = g\left(x^*, y^*, z^*\right) = h\left(x^*, y^*, z^*\right) = 0. \quad (2)$$

The equilibrium points can be classified according to the behavior of the trajectories of the system near them as follows:

- The equilibrium point e^* is (locally) *stable* if, for any $R > 0$, there is a $0 < r \leq R$ such that every trajectory within $B_r(e^*)$ at t_0, stay within $B_R(e^*)$ for every $t > t_0$. That is, e^* is stable if all solutions starting near e^* stay nearby.
- The equilibrium point e^* is (locally) unstable if it is not (locally) stable.

- The equilibrium point e^* is (locally) asymptotically stable if it is stable and, in addition, there is a $\delta > 0$ such that every trajectory within $B_\delta(e^*)$ at t approaches e^* as $t \to \infty$.

The following result characterizes the equilibrium points:

Theorem 1 *Let e^* be an equilibrium point of the system (1) such that $J(e^*)$ is the associated Jacobian matrix, and set λ_1, λ_2 and λ_3 its eigenvalues. Then the following hold:*

(1) e^* is (locally) asymptotically stable if $\mathrm{Re}(\lambda_i) < 0$ *for every* $1 \leq i \leq 3$.
(2) e^* is (locally) stable if $\mathrm{Re}(\lambda_i) \leq 0$ *for every* $1 \leq i \leq 3$.
(3) e^* is (locally) unstable if $\mathrm{Re}(\lambda_i) > 0$ *for some* $1 \leq i \leq 3$.

The *Routh–Hurwitz method* [4] helps to determine the position of the roots of the characteristic polynomial in the complex plane.

Theorem 2 *Let $P(\lambda) = \lambda^3 + p_1\lambda^2 + p_2\lambda + p_3$ be a polynomial whose coefficients are real and positive numbers. The necessary and sufficient conditions for all roots have real part negative are the following:*

$$\Delta_1 = p_1 > 0, \quad \Delta_2 = \begin{vmatrix} p_1 & p_3 \\ 1 & p_2 \end{vmatrix} > 0, \quad \Delta_3 = \begin{vmatrix} p_1 & p_3 & 0 \\ 1 & p_2 & 0 \\ 0 & p_1 & p_3 \end{vmatrix} > 0. \tag{3}$$

Corollary 1 *All roots of $P(\lambda) = \lambda^3 + p_1\lambda^2 + p_2\lambda + p_3$ have negative real parts if and only if $p_3 < p_1 p_2$.*

3 Description of the Proposed Mathematical Model

The mathematical model introduced in this work to study and simulate the spreading of a computer worm through a computer network is a compartmental model, that is, the population (of computers) can be classified into four types or compartments:

(1) *Susceptible* computers are those computers which have not been reached by the computer worm, and remain "healthy".
(2) *Infected* computers are those computers which have been reached by the computer worm and are able to transmit it to other computers. These infected computers can be further classified into the following two subtypes:

   a. *Carrier* computers: infected computers that the computer worm is not able to carry out its damaging function since the operating system of the computer does not match with the OS targeted by the malware.
   b. *Infectious* computers: infected computers whose OS is targeted by the computer worm, and consequently, the malware can carry out the payload.

(3) *Recovered* computers are those susceptible computers on which patches (or other necessary security software) have been installed in order to avoid their infection by the computer worm, or those infected computers (both carriers or infectious) that the malware has been successfully detected and removed.

Here, it is supposed that computers are endowed with the recovery state during a finite period of time. That is, the immunity period obtained when patches or other security software have been installed and ran is a temporary period. As a consequence, the recovery computers become susceptible again once the immunity period has finished. Moreover, as the propagation speed of computer worms is high, we can assume that the total population of computers remains constant through time.

3.1 The Equations that Govern the Dynamic of the Model

As was previously mentioned, the proposed model is a compartmental SCIRS model (see Fig. 1).

Since the population of computers remains constant over the time, then:

$$N = S(t) + C(t) + I(t) + R(t), \tag{4}$$

where N is the total number of computers, and $S(t)$, $C(t)$, $I(t)$ and $R(t)$ stand for the number of susceptible, carrier, infectious and recovered computers at time t, respectively. The dynamic of the model is governed by means of the following (autonomous) system of ordinary differential equations:

$$S'(t) = -a \cdot S(t) \cdot (I(t) + C(t)) - v \cdot S(t) + \varepsilon \cdot R(t), \tag{5}$$
$$C'(t) = a \cdot (1 - \delta)S(t) \cdot (I(t) + C(t)) - b \cdot C(t), \tag{6}$$
$$I'(t) = a \cdot \delta \cdot S(t)(I(t) + C(t)) - b \cdot I(t), \tag{7}$$
$$R'(t) = b \cdot (C(t) + I(t)) + v \cdot S(t) - \varepsilon \cdot R(t), \tag{8}$$

with the following initial conditions:

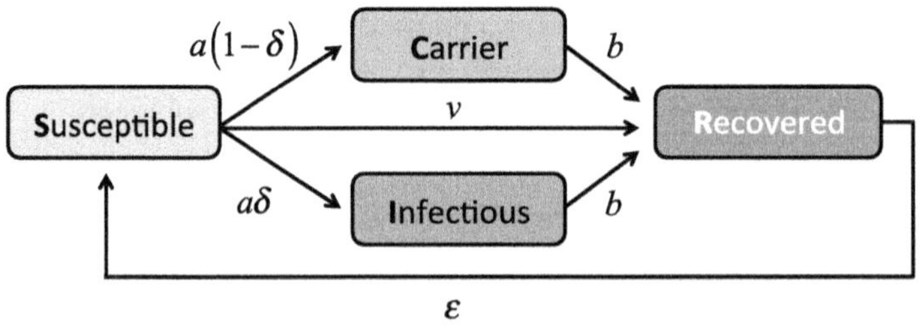

Fig. 1 Flow diagram representing the dynamic of the model

$$S(0) = S_0, C(0) = C_0, I(0) = I_0, R(0) = N - S_0 - C_0 - I_0, \tag{9}$$

$$S(t) \geq 0, C(t) \geq 0, I(t) \geq 0, R(t) \geq 0. \tag{10}$$

Furthermore, the parameters involved in this model are the following: the *transmission coefficient* a, the *vaccination coefficient* v, the *loss of immunity coefficient* ε, the fraction δ of computers whose operating system is the same as the attacked by the computer worm, and the *recovery coefficient* associated to infected computers b. Note that $0 \leq a, v, \varepsilon, \delta, b \leq 1$.

The Eq. (5) states that the variation of the number of susceptible computers is equal to the difference between the recovery computers that have lost the immunity $\varepsilon \cdot R(t)$, and the susceptible computers that have lost such status. The latter are the sum of the susceptible computers that become infected at every step of time (the *incidence*: $a \cdot S(t) \cdot (C(t) + I(t))$), plus the susceptible computers that are "vaccinated" at every step of time: $v \cdot S(t)$.

The Eq. (6) shows that the variation of the number of carrier computers is equal to the difference between the new susceptible computers (whose operating systems is different from the targeted one) that have been infected, $a \cdot (1 - \delta) \cdot S(t) \cdot (I(t) + C(t))$), and those carrier computers that have been recovered (once the computer worm has been detected and successfully deleted): $b \cdot C(t)$.

The evolution of the number of infectious computers is given in Eq. (7) such that the variation of this compartment is the different between the new susceptible computers (with the same operating system as the targeted by the malware) that have been infected, $a \cdot \delta \cdot S(t)(I(t) + C(t))$, and the infectious computers that have been recovered: $b \cdot I(t)$.

Finally, Eq. (8) shows that the variation of the number of recovered computers is the difference between the new recovered computers (both carriers and infectious), $b \cdot (C(t) + I(t))$ and the "vaccinated" susceptible computers: $v \cdot S(t)$, and the computers that loses the immunity: $\varepsilon \cdot R(t)$.

Note that the system of four ordinary differential equations (5)–(8) can be reduced to the following system of three ordinary differential equations by simply considering the Eq. (4):

$$S'(t) = -a \cdot S(t) \cdot (I(t) + C(t)) - v \cdot S(t) + \varepsilon \cdot R(t), \tag{11}$$

$$C'(t) = a \cdot (1 - \delta)S(t) \cdot (I(t) + C(t)) - b \cdot C(t), \tag{12}$$

$$I'(t) = a \cdot \delta \cdot S(t)(I(t) + C(t)) - b \cdot I(t). \tag{13}$$

3.2 Determination of Parameters

The parameters involved in the system of ordinary differential equations that governs the dynamic of the model are the transmission coefficient a, the "vaccination" coefficient v, the immunity coefficient ε, the fraction of population running under the same operating system as the targeted by the malware δ, and the recovery coefficient

b. In what follows, we will describe how to determine theoretically each of these coefficients.

3.2.1 The Transmission Coefficient a

The incidence is defined as the number of new infected computers that appeared in each step of time. Mathematically, it is defined as $\lambda \cdot S(t)$, where λ is the force of infection. To simulate the spreading of computer worms it is very important to determine correctly λ; this can be achieved in different ways taking into account the different choices we can make to estimate the number of contacts between the computers.

The malware infection is transmitted through adequate and effective contacts (infectious contacts) between susceptible and infectious computers. Note that a contact is said to be adequate when it enables the transmission of malware; moreover, an adequate contact is said to be effective (and, consequently, an infectious contact) when the malware successfully reaches the host computer. In this work, it is supposed that the transmission vector is defined by the emails and, as a consequence, the infectious contacts will be made by sending the malicious code by email.

The number of times a computer comes into adequate contact with other computer per unit time is defined as the contact rate. Usually, the contact rate depends on the total number of computers N, and consequently $k = k(N)$. Let q be the probability that an adequate contact becomes an infectious contact, then $q \cdot k(N)$ stands for the total number of infectious contacts between each computer with the rest of computers per unit time. Consequently, the force of infection is given by the following equation:

$$\lambda = q \cdot \frac{k(N)}{N} \cdot (C(t) + I(t)). \tag{14}$$

Note that this coefficient depends on time (it is not constant through the duration of the epidemic period).

The epidemiological significance of this coefficient is as follows: as $\frac{k(N)}{N}$ is the average of the number of adequate contacts between each computer and the rest of computers of the network at every step of time, then $q \cdot \frac{k(N)}{N}$ is the average number of infectious contacts of each susceptible computer with the rest of computers of the network at every step of time. Consequently $\lambda = q \cdot \frac{k(N)}{N} \cdot (P(t) + I(t))$ is the average of infectious contacts of each susceptible computer with the total number of infected computers at every step of time.

As a consequence, the transmission coefficient can be defined as follows:

$$a = q \cdot \frac{k(N)}{N}. \tag{15}$$

As mentioned above, the degree of plausibility of the simulations obtained with the mathematical model depends strongly on the choice of the transmission coefficient and more specifically, on the contact rate $k(N)$. In this sense, several explicit expressions for the contact rate can be considered, and we can highlight the following:

- *Bilinear* contact rate: $k(N) = \alpha \cdot N$, where α stands for the average of contacts between two computers of the network at every step of time. It yields to the so-called *bilinear incidence* or *mass action*.
- *Standard* contact rate: $k(N) = \delta$, where δ stands the average of contacts of each computer with the rest of computers of the network at every step of time. Consequently, the *standard incidence* is obtained.

In this work, the proposed model is based on the bilinear incidence, then: $\lambda \cdot S(t) = a \cdot S(t) \cdot (P(t) + I(t))$, where $a = q \cdot \alpha$.

3.2.2 The "Vaccination" Coefficient v

The susceptible computers can acquire temporary immunity when necessary software patches, operating system updates or/and other security software are installed. Then $v = \nu \cdot \xi$, where ν stands for the fraction of the susceptible population that has been vaccinated at every step of time, and $0 \leq \xi \leq 1$ is the success rate of vaccination.

3.2.3 The Immunity Coefficient ε

The immunity period (whose length is denoted by T_I) is the period of time between the instant at which the computer worm is removed from the infected computer and the instant when the computer becomes susceptible again. Suppose that at a particular step of time (for example, $t = 0$, without loss of generality) all computers are isolated (and, in particular, the recovered), then the evolution of this compartment is given by the following equation:

$$\frac{dR}{dt} = -\varepsilon \cdot R(t), \tag{16}$$

whose solution is:

$$R(t) = R(0) \cdot e^{-\varepsilon \cdot t}, \tag{17}$$

that is $e^{-\varepsilon \cdot t} = \frac{R(t)}{R(0)}$ is the fraction of the total number of computers that are still being recovered t time units after isolation.

Consequently, we can suppose that $e^{-\varepsilon \cdot t}$ is an estimator of the probability to remain susceptible t time units after isolation. Thus $1 - e^{-\varepsilon \cdot t}$ stands for the fraction of recovery computers that ceases to be at time step t, that is, it can be supposed that it estimates the probability to stop being recovered at time t. As a consequence $1 - e^{-\varepsilon \cdot t}$ estimates the probability that the length of the immunity period will be t.

Since $\mathscr{F}_R(t) = 1 - e^{-\varepsilon \cdot t}$ is a strictly increasing function of class $\mathscr{C}^\infty(\mathbb{R})$ then it is the probability distribution function associated to the random variable X_R representing the length of the immunity period. Thus, $\mathscr{F}_R(t) = p(X_R \leq t)$. Moreover, the associated probability density function is $f_R(t) = \mathscr{F}'_R(t) = \varepsilon \cdot e^{-\varepsilon \cdot t}$, so that the expected value of X_R is the mathematical expectation:

$$\mathscr{E}[X_R] = \int_0^\infty t \cdot f_R(t)dt = \int_0^\infty \left(\varepsilon \cdot t \cdot e^{-\varepsilon \cdot t}\right) dt = \frac{1}{\varepsilon}. \tag{18}$$

As a consequence $\varepsilon = \frac{1}{T_I}$.

3.2.4 The Recovery Coefficient b

As discussed above, the recovery of an infected computer depends on two factors: the malware detection and the malware successful removal (once it has been detected). As a consequence if d stands for the probability to detect the computer worm at every step of time, and e is the probability of successful removal, then $b = d \cdot e$.

3.2.5 The Coefficient δ

Usually, the targeted computers of malware are those whose operative system is one in particular, such that no harm is caused to the rest of computers (even if they are infected and act as transmission vectors). In this sense δ stands for the fraction of the number of computers running under the targeted operating system.

3.3 *The Basic Reproductive Number*

The basic reproductive number, $\mathscr{R}_0$, is perhaps the most important parameter in the study of malware propagation since its numerical value will indicate whether or not an outbreak of a malicious code will become epidemic (the number of infected computers will grow). Roughly speaking the basic reproductive number can be defined as the average number of secondary infections that occur when only one infectious computer appears in a completely susceptible host population of computers. Then if $\mathscr{R}_0 > 1$ the number of infected computers increases (and the outbreak becomes epidemic), whereas if $\mathscr{R}_0 \leq 1$ the computer worm will not spread. Consequently, it is very important to explicitly determine this threshold parameter.

It is possible to compute the $\mathscr{R}_0$ from the system of ordinary differential equations (5)–(8) and it is therefore necessary to compute the equilibrium points of such system. These points are the solutions of the following nonlinear system:

$$0 = -a \cdot S \cdot (I + C) - v \cdot S + \varepsilon \cdot R,$$
$$0 = a \cdot (1 - \delta)S \cdot (I + C) - b \cdot C,$$

$$0 = a \cdot \delta \cdot S(I + C) - b \cdot I,$$
$$0 = b \cdot C + b \cdot I + v \cdot S - \varepsilon \cdot R.$$

A simple computation shows that this system has two solutions: the disease-free equilibrium point $E_0^* = \left(S_0^*, C_0^*, I_0^*, R_0^*\right) = \left(\frac{\varepsilon N}{\varepsilon + v}, 0, 0, \frac{vN}{\varepsilon + v}\right)$, and the endemic equilibrium point: $E_1^* = \left(S_1^*, C_1^*, I_1^*, R_1^*\right)$, where:

$$S_1^* = \frac{b}{a}, \quad C_1^* = -\frac{(1-\delta)\,(b\,(v+\varepsilon) - \varepsilon a N)}{a\,(\varepsilon + b)} \tag{19}$$

$$I_1^* = -\frac{\delta\,(b\,(v+\varepsilon) - \varepsilon a N)}{a\,(\varepsilon + b)}, \quad R_1^* = \frac{b\,(aN - b + v)}{a\,(\varepsilon + b)}. \tag{20}$$

Note that the endemic equilibrium point exists if $b\,(v + \varepsilon) - \varepsilon a N < 0$, that is, the total number of computers exceeds a certain threshold value:

$$N > \frac{b\,(v+\varepsilon)}{\varepsilon a}. \tag{21}$$

The *next generation method* [1] will be used to determine the basic reproductive number. This is a general method of deriving $\mathscr{R}_0$ when more than one class of infected computers are considered (recall that in our case we take into account two subtypes: carriers and infectious). If

$$F_C\,(S, C, I, R) = a \cdot (1-\delta) \cdot S \cdot (I + C)\,, \tag{22}$$
$$F_I\,(S, C, I, R) = a \cdot \delta \cdot S \cdot (I + C)\,, \tag{23}$$

are the appearance functions of new carrier and infectious computers respectively, then the appearance matrix F is defined as follows:

$$F = \begin{pmatrix} \frac{\partial F_C}{\partial C} & \frac{\partial F_C}{\partial I} \\ \frac{\partial F_I}{\partial C} & \frac{\partial F_I}{\partial I} \end{pmatrix} = \begin{pmatrix} a\,(1-\delta)\,S & a\,(1-\delta)\,S \\ a\delta S & a\delta S \end{pmatrix}. \tag{24}$$

On the other hand, if V_C^+, V_C^- (*resp.* V_I^+, V_I^-) are the transfer functions associated to carrier computers (*resp.* infectious computers), then $V_C^+\,(S, C, I, R) = V_I^+\,(S, C, I, R) = 0$, $V_C^-\,(S, C, I, R) = bC$, $V_I^-\,(S, C, I, R) = bI$, and the following is obtained:

$$V_C\,(S, C, I, R) = V_C^-\,(S, C, I, R) - V_C^+\,(S, C, I, R) = bC, \tag{25}$$
$$V_I\,(S, C, I, R) = V_I^+\,(S, C, I, R) - V_I^+\,(S, C, I, R) = bI. \tag{26}$$

As a consequence:

$$V=\begin{pmatrix}\frac{\partial V_C}{\partial C} & \frac{\partial V_C}{\partial I}\\ \frac{\partial V_I}{\partial C} & \frac{\partial V_I}{\partial I}\end{pmatrix}=\begin{pmatrix}b & 0\\ 0 & b\end{pmatrix}. \tag{27}$$

Therefore the basic reproductive number $\mathscr{R}_0$ is the spectral radius of the next generation matrix in the disease-free equilibrium:

$$\left(F\cdot V^{-1}\right)_{E_0^*}=\begin{pmatrix}\frac{a\left(1-\delta\right)\varepsilon N}{b\left(\varepsilon+v\right)} & \frac{a\left(1-\delta\right)\varepsilon N}{b\left(\varepsilon+v\right)}\\ \frac{a\delta\varepsilon N}{b\left(\varepsilon+v\right)} & \frac{a\delta\varepsilon N}{b\left(\varepsilon+v\right)}\end{pmatrix}, \tag{28}$$

that is:

$$\mathscr{R}_0=\frac{a\varepsilon N}{b\left(\varepsilon+v\right)}. \tag{29}$$

Note that in order to prevent that an epidemic occurs it is mandatory to reduce the basic reproductive number $\mathscr{R}_0$ as necessary. In this sense, and taking into account its explicit expression given by Eq. (29), this is achieved by considering some of the following measures: (1) Reducing the total number of computers on the network N by means of, for example, isolation. (2) Reducing the transmission coefficient a by reducing the number of effective contacts between computers or extreme caution when opening suspicious emails. (3) Increasing the recovery rate b by improving the performance of antivirus software.

3.4 Local Stability of the Disease-Free Equilibrium

The following result holds:

Theorem 3 *The disease-free equilibrium point* $E_0^*=\left(\frac{\varepsilon N}{\varepsilon+v},0,0,\frac{vN}{\varepsilon+v}\right)$ *is locally asymptotically stable if and only if* $\mathscr{R}_0<1$.

Proof The Jacobian matrix associated to the system of ordinary differential equations (11)–(13) in the disease-free equilibrium is:

$$J\left(E_0^*\right)=\begin{pmatrix}-v-\varepsilon & -\frac{\varepsilon aN}{\varepsilon+v}-\varepsilon & -\frac{\varepsilon aN}{\varepsilon+v}-\varepsilon\\ 0 & -b+\frac{\varepsilon\left(1-\delta\right)aN}{\varepsilon+v} & \frac{\varepsilon\left(1-\delta\right)aN}{\varepsilon+v}\\ 0 & \frac{\varepsilon\delta aN}{\varepsilon+v} & -b+\frac{\varepsilon\delta aN}{\varepsilon+v}\end{pmatrix}, \tag{30}$$

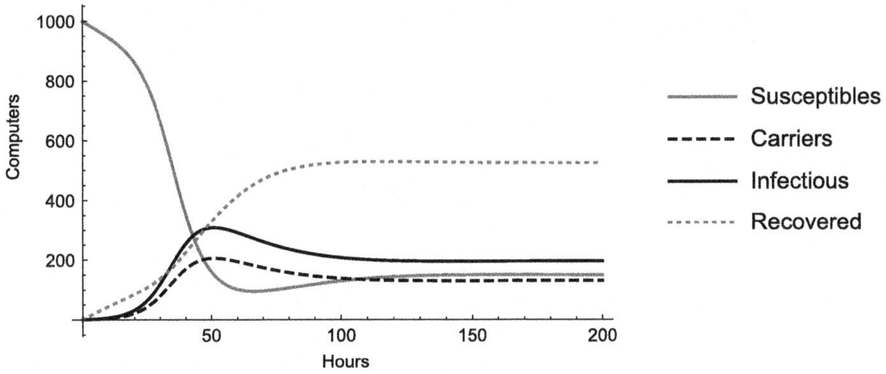

Fig. 2 Evolution of the compartments when $\mathscr{R}_0 < 1$

so that a simple calculus shows that its eigenvalues are:

$$\lambda_1 = -\nu - \varepsilon, \quad \lambda_2 = -b, \quad \lambda_3 = -\frac{\varepsilon a N - (\varepsilon + \nu)\, b}{\varepsilon + \nu} = b\,(-1 + \mathscr{R}_0)\,. \tag{31}$$

As a consequence $\mathrm{Re}\,(\lambda_1) < 0$ and $\mathrm{Re}\,(\lambda_2) < 0$. Furthermore, $\mathrm{Re}\,(\lambda_3) < 0$ iff $\mathscr{R}_0 < 1$, thus finishing.

The evolution of the different compartments when $a = 0.000138$ ($\alpha = \frac{1}{72}, q = 0.01$), $\nu = 0.005$ ($\upsilon = \frac{1}{100}, \xi = 0.5$), $\varepsilon = \frac{1}{48}$, $b = 0.0375$ ($d = \frac{1}{24}, e = 0.9$) and $\delta = 0.6$ is shown in Fig. 2. The time is measured in hours, $S\,(0) = 100$, $C\,(0) = 1$, $I\,(0) = 1$, $R\,(0) = 0$ and the simulation period comprises the first 150 h after the onset of the first infectious computer. In this case $\mathscr{R}_0 \approx 0.304659 < 1$ and consequently the number of infected computers does not increase.

The disease-free equilibrium is $E_0^* \approx (82.2581, 0, 0, 19.7419)$, so that at $t = 150$ the value of the compartments are $S\,(150) \approx 81.7380$, $C\,(150) \approx 0.0073$, $I\,(150) \approx 0.0148$ and $R\,(150) \approx 19.2399$.

3.5 Local Stability of the Endemic Equilibrium

As was previously mentioned, the model exhibit the endemic equilibrium point $E_1^* = \left(S_1^*, P_1^*, I_1^*, R_1^*\right)$ given by (19) and (20) if $\Omega = b\,(\varepsilon + \nu) - \varepsilon a N < 0$. In this case, the following result holds:

Theorem 4 *The endemic equilibrium* $E_1^* = \left(S_1^*, C_1^*, I_1^*, R_1^*\right)$ *defined by (19) and (20) is locally asymptotically stable if* $\mathscr{R}_0 > 1$.

Proof The Jacobian matrix of the system in the endemic equilibrium point E_1^* is:

$$J\left(E_1^*\right)=\begin{pmatrix} \dfrac{\varepsilon\left(b-v-aN\right)}{\varepsilon+b}-\varepsilon & -b-\varepsilon & -b-\varepsilon \\ -\dfrac{(1-\delta)\left((\varepsilon+v)\,b-\varepsilon aN\right)}{\varepsilon+b} & -\delta b & (1-\delta)\,b \\ -\dfrac{\delta\left((\varepsilon+v)\,b-\varepsilon aN\right)}{\varepsilon+b} & \delta b & -(1-\delta)\,b \end{pmatrix}, \tag{32}$$

such that its characteristic polynomial is the following:

$$\begin{aligned} p(\lambda) &= -\lambda^3+\frac{-aN\varepsilon-b^2-b\varepsilon-v\varepsilon-\varepsilon^2}{b+\varepsilon}\lambda^2 \\ &+\frac{b\varepsilon(-aN+v+\varepsilon)-abN\varepsilon-aN\varepsilon^2+b^2(v+\varepsilon)-bv\varepsilon-b\varepsilon^2}{b+\varepsilon}\lambda \\ &+\frac{b^2\varepsilon(-aN+v+\varepsilon)-abN\varepsilon^2+b^3(v+\varepsilon)}{b+\varepsilon}. \end{aligned} \tag{33}$$

By applying the Routh–Hurwitz stability criterion, the real part of the eigenvalues of $p(\lambda)$ will be negative when the following conditions hold:

$$\Delta_1=\frac{b^2+bv+2b\varepsilon+v\varepsilon-\Omega+\varepsilon^2}{b+\varepsilon}>0, \tag{34}$$

$$\Delta_2=\begin{vmatrix} \Delta_1 & -b\Omega \\ 1 & \frac{a\varepsilon N}{\mathscr{R}_0}-\frac{2b+\varepsilon}{b+\varepsilon}\Omega \end{vmatrix}>0, \tag{35}$$

$$\Delta_3=\begin{vmatrix} \Delta_1 & -b\Omega & 0 \\ 1 & \frac{a\varepsilon N}{\mathscr{R}_0}-\frac{2b+\varepsilon}{b+\varepsilon}\Omega & 0 \\ 0 & \Delta_1 & -b\Omega \end{vmatrix}=-b\Omega\Delta_2>0. \tag{36}$$

Since $\Omega<0$, then $\Delta_1>0$. Moreover, $\Delta_3>0$ if $\Delta_2>0$. A simple (and long) computation shows that this inequality holds if $\mathscr{R}_0>1$. Then the endemic equilibrium is locally asymptotically stable if $\mathscr{R}_0>1$.

The evolution of the different classes of computers when $a=0.000208$ ($\alpha=\frac{1}{48}, q=0.01$), $v=0.005$ ($\nu=\frac{1}{100}, \xi=0.5$), $\varepsilon=\frac{1}{48}$, $b=0.03125$ ($d=\frac{1}{24}, e=0.75$) and $\delta=0.6$ is shown in Fig. 3. It is suppose that $S(0)=1000$, $C(0)=1$, $I(0)=1$, $R(0)=0$. Moreover, the time is measured in hours and the simulation period represents the first 200 h after the appearance of the first infectious computers. In this simulation $\mathscr{R}_0\approx 5.3871>1$ and consequently the outbreak becomes epidemic. In this case the endemic equilibrium is $E_1^*=(150, 130.56, 195.84, 525.6)$, and the values of the different compartments at $t=200$ are the following $S(200)\approx 150.166$, $P(200)\approx 130.524$, $I(200)\approx 195.785$ and $R(200)\approx 525.525$.

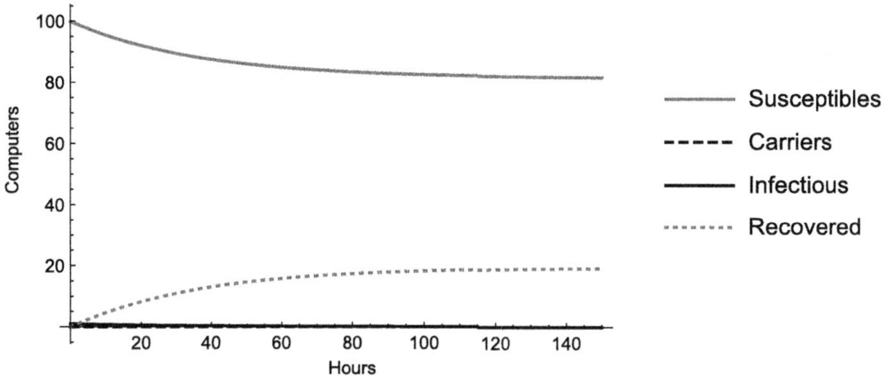

Fig. 3 Temporal evolution of the compartments when $\mathscr{R}_0 > 1$

4 Conclusion

The great majority of mathematical models proposed to date whose goal is to simulate malware spreading are based on systems of ordinary differential equations. The stability theory plays an important role as it yields to the classification of the equilibrium points taking into account the behavior of the trajectories (and, consequently, the evolution of the different types of computers). This classification depends on the so-called basic reproductive number, $\mathscr{R}_0$. That is, if $\mathscr{R}_0 < 1$ (an outbreak does not become epidemic) the disease-free equilibrium (where there are not infected computers) is locally and asymptotically stable, whereas the endemic equilibrium (there are always infected computers) point is locally unstable. On the other hand, if $\mathscr{R}_0 > 1$ (the number of infected computers increases) then the disease-free equilibrium point is unstable whereas the endemic equilibrium point is locally and asymptotically stable.

Acknowledgments This work has been supported by MINECO (Spain) and European FEDER Fund under the grant MoMaCIS (TIN2014-55325-C2-2-R), and by Junta de Castilla y León (Spain).

References

1. Diekmann, O., Heesterbeek, J.A.P., Metz, J.A.J.: On the definition and the computation of the basic reproduction ratio R_0 in models for infectious diseases in heterogeneous populations. J. Math. Biol. **28**(4), 365–382 (1990)
2. Jang-Jaccard, J., Nepal, S.: A survey of emerging threats in cybersecurity. J. Comput. Syst. Sci. **80**, 973–993 (2014)
3. Martín del Rey, A.: Mathematical modeling of the propagation of malware: a review. Secur. Commun. Netw. **8**, 2561–2579 (2015)
4. Merkin, D.R.: Introduction to the Theory of Stability. Texts in Applied Mathemtics, vol. 24. Springer, New York (1997)

On the Non-triviality of the Eight-Form $\tau_4(\omega)$ on Manifolds with a Spin(9)-Structure

I.V. Mykytyuk

Dedicated to Jaime Muñoz Masqué on the occasion of his 65th birthday.

Abstract It is proved that Parton-Piccinni's expression $\tau_4(\omega)$ of the canonical 8-form on a manifold with holonomy group Spin(9) is not trivial, by using the properties of the octonions.

Keywords Spin(9) holonomy · Canonical 8-form

1 Introduction and Preliminaries

The group Spin(9) belongs to Berger's list of restricted holonomy groups of locally irreducible Riemannian manifolds which are not locally symmetric. Manifolds with holonomy group Spin(9) have been studied by Alekseevsky [3], Brown and Gray [6], Friedrich [8, 9], and Lam [10], among other authors. As proved in [3, 6], a connected, simply-connected, complete non-flat Spin(9)-manifold is isometric to either the Cayley projective plane $\mathbb{O}P(2) \cong F_4/\mathrm{Spin}(9)$ or its dual symmetric space, the Cayley hyperbolic plane $\mathbb{O}H(2) \cong F_{4(-20)}/\mathrm{Spin}(9)$.

Moreover, Δ_9 being the unique irreducible 16-dimensional Spin(9)-module, the Spin(9)-module $\Lambda^8(\Delta_9^*)$ contains one and only one (up to a non-zero factor)

I.V. Mykytyuk (✉)
Institute of Mathematics, Cracow University of Technology, Warszawska 24, 31155 Cracow, Poland
e-mail: mykytyuk_i@yahoo.com

I.V. Mykytyuk
Institute of Applied Problems of Mathematics and Mechanics, Naukova Str. 3b, Lviv 79601, Ukraine

M. Castrillón López et al. (eds.), *Geometry, Algebra and Applications: From Mechanics to Cryptography*, Springer Proceedings in Mathematics & Statistics 161, DOI 10.1007/978-3-319-32085-4_15

8-form Ω_0^8 which is Spin(9)-invariant and defines the unique parallel form on $\mathbb{O}P(2)$. It induces a *canonical 8-form* Ω^8 on any 16-dimensional manifold with a fixed Spin(9)-structure. This form is said to be canonical because (cf. [6, p. 48]) it yields, for the compact case, a generator of $H^8(\mathbb{O}P(2), \mathbb{R})$.

Some explicit expressions of Ω_0^8 have been given. The first one by Brown and Gray in [6, p. 49] in terms of a Haar integral. Other expression was then given by Brada and Pécaut-Tison [4, pp. 150,153], [5], by using a "cross product." Unfortunately, their formula is not correct (see [7] for more detailed explanations). Another expression was then given by Abe and Matsubara in [2, p. 8] as a sum of 702 suitable terms (see also Abe [1]). Their formula contains some errors (see [7] for more detailed explanations).

In the paper [7] Castrillón López, Gadea and Mykytyuk found an explicit expression for the canonical 8-form Ω^8 on a Spin(9)-manifold in terms of the 9×9 skew-symmetric matrix $\omega = (\omega_{ij})$ of the involved local Kähler 2-forms. They proved the invariance and non-triviality of Ω^8 using the properties of the automorphisms of the octonion algebra.

Later, another expression $\tau_4(\omega)$ for the canonical 8-form, as the fourth coefficient of the characteristic polynomial of the matrix ω, was proposed by Parton and Piccinni in [11]. To prove the non-triviality of $\tau_4(\omega)$ they performed computer computations with the help of the software Mathematica.

We recall that a Spin(9)-structure on a connected, oriented 16-dimensional Riemannian manifold (M, g) is defined as a reduction of its bundle of oriented orthonormal frames $SO(M)$, via the spin representation $\rho(\mathrm{Spin}(9)) \subset SO(16)$. Equivalently (Friedrich [8, 9]), a Spin(9)-structure is given by a nine-dimensional subbundle ν^9 of the bundle of endomorphisms $\mathrm{End}(TM)$ locally spanned by $I_i \in \Gamma(\nu^9)$, $0 \leqslant i \leqslant 8$, satisfying the relations $I_iI_j + I_jI_i = 0$, $i \neq j$, $I_i^2 = \mathrm{I}$, $I_i^T = I_i$, $\mathrm{tr}\, I_i = 0$, $i, j = 0, \ldots, 8$. These endomorphisms define 2-forms ω_{ij}, $0 \leqslant i < j \leqslant 8$, on M locally by $\omega_{ij}(X, Y) = g(X, I_iI_jY)$. Similarly, using the skew-symmetric involutions $I_iI_jI_k$, $0 \leqslant i < j < k \leqslant 8$, one can define 2-forms σ_{ijk}. The 2-forms $\{\omega_{ij}, \sigma_{ijk}\}$ are linearly independent and a local basis of the bundle Λ^2M.

The expression for the (global) canonical 8-form on the Spin(9)-manifold (M, g, ν^9) is given (see [7, Theorem1]) by

$$\Omega^8 = \sum_{\substack{i,j=0,\ldots,8\\ i',j'=0,\ldots,8}} \omega_{ij} \wedge \omega_{ij'} \wedge \omega_{i'j} \wedge \omega_{i'j'}, \tag{1}$$

where $\omega_{ij} = -\omega_{ji}$ if $i > j$ and $\omega_{ij} = 0$ if $i = j$. The expression $\tau_4(\omega)$ (see [11]) for the canonical 8-form on (M, g, ν^9) is given by

$$\tau_4(\omega) = \sum_{0\leqslant\alpha_1<\alpha_2<\alpha_3<\alpha_4\leqslant 8} (\omega_{\alpha_1\alpha_2} \wedge \omega_{\alpha_3\alpha_4} - \omega_{\alpha_1\alpha_3} \wedge \omega_{\alpha_2\alpha_4} + \omega_{\alpha_1\alpha_4} \wedge \omega_{\alpha_2\alpha_3})^2. \tag{2}$$

Remark that the fourth coefficient $\tau_4(\omega)$ (above) of the characteristic polynomial of the skew-symmetric matrix ω is computed with a summation over the squared Pfaffians of the principal 4×4-submatrices of ω.

The main purpose of the present paper is to prove, using the properties of the automorphisms of the octonion algebra (and without computer calculations), the next result.

Proposition 1 *The* Spin(9)*-invariant* 8*-form* $\tau_4(\omega)$ *is not trivial.*

As a consequence, one has the following corollary.

Corollary 1 *The two expressions* Ω^8 *and* $\tau_4(\omega)$ *of the canonical* 8*-form on the* Spin(9)*-manifold* (M, g, ν^9) *are related by* $\Omega^8 = -4\tau_4(\omega)$.

2 Proof of Proposition 1

To prove that the form $\tau_4(\omega)$ is nontrivial consider its restriction $\tau_4(\omega)|T_pM$ and the restrictions of the 2-forms $\omega_{ij}|T_pM$, $i, j = 0, \ldots, 8$, where $p \in M$ is an arbitrarily fixed point. To simplify the notation we will write $\tau_4(\omega)$ and ω_{ij} for these restrictions.

There exists an isomorphism between $\mathbb{O}^2 \equiv \mathbb{R}^{16}$ and T_pM such that the restriction of g at $p \in M$ induces the standard scalar product $\langle \cdot, \cdot \rangle$ of $\mathbb{O}^2$, given by

$$\langle (x_1, x_2), (y_1, y_2) \rangle = \langle x_1, y_1 \rangle + \langle x_2, y_2 \rangle, \quad \langle x_a, y_a \rangle = \frac{1}{2}(x_a \bar{y}_a + y_a \bar{x}_a), \tag{3}$$

for $a = 1, 2$. Since $\tau_4(\omega)$ is an 8-form, we consider the eight vectors $X_i = (u_i, 0)$, where $u_0 = 1 \in \mathbb{O}$ and u_i, $i = 1, \ldots, 7$, stand for the imaginary units of $\mathbb{O}$, and two vectors $X = (x, 0)$ and $Y = (y, 0)$ belonging to the space $\mathbb{O}^2$. Then for $i, j = 0, \ldots, 7$, $i \neq j$, we have [7, Eq. 7]

$$\omega_{ij}(X, Y) = \langle x, u_i(\bar{u}_j y) \rangle \qquad \text{and} \qquad \omega_{i8}(X, Y) = 0. \tag{4}$$

We can rewrite the expression for $\omega_{ij}(X, Y)$ as

$$\omega_{ij}(X, Y) = \langle x, u_i(\bar{u}_j y) \rangle = \langle \bar{u}_i x, \bar{u}_j y \rangle = \langle \bar{x} u_i, \bar{y} u_j \rangle, \tag{5}$$

because (cf. [6, Sect. 2]) for arbitrary octonions $a, b, c \in \mathbb{O}$, one has

$$\langle ab, c \rangle = \langle b, \bar{a}c \rangle = \langle a, c\bar{b} \rangle \ \text{ and } \ \langle a, b \rangle = \langle \bar{a}, \bar{b} \rangle. \tag{6}$$

It is clear that the 8-form $\tau_4(\omega)$ is nontrivial if

$$\begin{aligned} &\tau(\alpha_1, \alpha_2, \alpha_3, \alpha_4) \\ &\quad = (\omega_{\alpha_1\alpha_2} \wedge \omega_{\alpha_3\alpha_4} - \omega_{\alpha_1\alpha_3} \wedge \omega_{\alpha_2\alpha_4} + \omega_{\alpha_1\alpha_4} \wedge \omega_{\alpha_2\alpha_3})^2 (X_0, \ldots, X_7) > 0 \end{aligned}$$

for any subset $\{\alpha_1, \alpha_2, \alpha_3, \alpha_4\} \subset \{0, \ldots, 7\}$.

To reduce calculations we will prove that, for an arbitrary subset $\{\alpha_1, \alpha_2, \alpha_3, \alpha_4\}$ of $\{0, \ldots, 7\}$, we have that

$$\text{either } \tau(\alpha_1, \alpha_2, \alpha_3, \alpha_4) = \tau(0, 1, 2, 3) \text{ or } \tau(\alpha_1, \alpha_2, \alpha_3, \alpha_4) = \tau(0, 1, 2, 4). \tag{7}$$

Let S_4^α be the permutation group acting on the set $\{\alpha_1, \alpha_2, \alpha_3, \alpha_4\} \subset \{0, \ldots, 7\}$. The form $\tau_4(\omega)$ is a linear combination of the 8-forms $V(\alpha_1, \alpha_2, \alpha_3, \alpha_4; \beta^\alpha, \gamma^\alpha)$, where $\beta^\alpha, \gamma^\alpha \in S_4^\alpha$, $\beta^\alpha = (\beta_1, \beta_2, \beta_3, \beta_4)$, $\gamma^\alpha = (\gamma_1, \gamma_2, \gamma_3, \gamma_4)$ and

$$V(\alpha_1, \alpha_2, \alpha_3, \alpha_4; \beta^\alpha, \gamma^\alpha) = \omega_{\beta_1\beta_2} \wedge \omega_{\beta_3\beta_4} \wedge \omega_{\gamma_1\gamma_2} \wedge \omega_{\gamma_3\gamma_4}.$$

Let S_8 be the permutation group acting on the set $B = \{u_0, \ldots, u_7\}$ and let $B^\pm = \{\pm u_0, \ldots, \pm u_7\}$. Let $\{w_1, w_2, w_4, w_4\}$ be an arbitrary subset of the set $B^\pm$ such that $w_i \neq \pm w_j$ for all $i \neq j$. Denote by S_4 the permutation group acting on the set $\{1, 2, 3, 4\}$. Choose two permutations $\beta, \gamma \in S_4$, $\beta = (\beta_1, \beta_2, \beta_3, \beta_4)$ and $\gamma = (\gamma_1, \gamma_2, \gamma_3, \gamma_4)$. Put

$$V(w_1, w_2, w_3, w_4; \beta, \gamma) = 2^{-4} \sum_{\sigma \in S_8} C_\sigma(w_1, w_2, w_3, w_4; \beta, \gamma),$$

where $C_\sigma(w_1, w_2, w_3, w_4; \beta, \gamma)$, for $\sigma = (u_{i_0}, \ldots, u_{i_7})$, is given by

$$\begin{aligned} &C_\sigma(w_1, w_2, w_3, w_4; \beta, \gamma) \\ &= \varepsilon(\sigma)\langle u_{i_0}, w_{\beta_1}(\bar{w}_{\beta_2}u_{i_1})\rangle\langle u_{i_2}, w_{\beta_3}(\bar{w}_{\beta_4}u_{i_3})\rangle\langle u_{i_4}, w_{\gamma_1}(\bar{w}_{\gamma_2}u_{i_5})\rangle\langle u_{i_6}, w_{\gamma_3}(\bar{w}_{\gamma_4}u_{i_7})\rangle. \end{aligned}$$

As the elements w_1, w_2, w_3, w_4 occur in this expression twice, we have

$$V(w_1, w_2, w_3, w_4; \beta, \gamma) = V(\pm w_1, \pm w_2, \pm w_3, \pm w_4; \beta, \gamma). \tag{8}$$

By definition $V(\alpha_1, \alpha_2, \alpha_3, \alpha_4; \beta^\alpha, \gamma^\alpha) = V(u_{\alpha_1}, u_{\alpha_2}, u_{\alpha_3}, u_{\alpha_4}; \beta, \gamma)$, where the permutation group S_4^α is identified naturally with the group S_4. We will write simply $V(w_1, w_2, w_3, w_4)$ and $C_\sigma(w_1, w_2, w_3, w_4)$ if $\beta = \gamma = \mathrm{id}$.

We now prove two lemmas.

Lemma 1 *For an arbitrary automorphism Φ of the algebra $\mathbb{O}$ preserving the set $B^\pm$, one has $V(w_1, w_2, w_3, w_4; \beta, \gamma) = V(\Phi(w_1), \Phi(w_2), \Phi(w_3), \Phi(w_4); \beta, \gamma)$.*

Proof It is clear that $\Phi(u_k) = \varepsilon_{u_k}^\Phi \sigma^\Phi(u_k)$, where $\varepsilon_{u_k}^\Phi = \pm 1$ and σ^Φ is some permutation in S_8. Moreover, since Φ is an element of the compact exceptional Lie group $G_2 \subset SO(7)$, we have $\prod_{k=0}^7 \varepsilon_{u_k}^\Phi \cdot \varepsilon(\sigma^\Phi) = 1$ and, consequently, we have $C_{\sigma^\Phi\sigma}(\Phi(w_1), \Phi(w_2), \Phi(w_3), \Phi(w_4); \beta, \gamma) = C_\sigma(w_1, w_2, w_3, w_4; \beta, \gamma)$, because $\varepsilon(\sigma^\Phi\sigma) = \varepsilon(\sigma^\Phi)\varepsilon(\sigma)$ and $\sigma^\Phi\sigma(u_k) = \varepsilon_{\sigma(u_k)}^\Phi \Phi(\sigma(u_k))$ (Φ preserves the scalar product and commutes with the conjugation). Since $\sigma^\Phi S_8 = S_8$, we conclude. □

Lemma 2 *For any $u \in B^{\pm}$, one has*

$$V(w_1, w_2, w_3, w_4; \beta, \gamma) = V(uw_1, uw_2, uw_3, uw_4; \beta, \gamma).$$

Proof Since the lemma is obvious for $u = \pm u_0$, assume that $u \neq \pm u_0$. Due to the relations (5), we can rewrite the expression for $C_\sigma(w_1, w_2, w_3, w_4; \beta, \gamma)$ as

$$\varepsilon(\sigma)\langle \bar{w}_{\beta_1} u_{i_0}, \bar{w}_{\beta_2} u_{i_1}\rangle\langle \bar{w}_{\beta_3} u_{i_2}, \bar{w}_{\beta_4} u_{i_3}\rangle\langle \bar{w}_{\gamma_1} u_{i_4}, \bar{w}_{\gamma_2} u_{i_5}\rangle\langle \bar{w}_{\gamma_3} u_{i_6}, \bar{w}_{\gamma_4} u_{i_7}\rangle.$$

But for arbitrary octonions a, b, c, their associator $(a, b, c) = (ab)c - a(bc)$ is skew-symmetric with respect to the second and third arguments, i.e. $(ab)c + (ac)b = a(bc + cb)$ (cf. [6, Sect. 2]). Thus, if $u_k u = -uu_k$ then $(au_k)u = (-au)u_k$. Since $u \neq \pm u_0$, one has $u_k u \neq -uu_k$ if and only if either $u_k = u_0$ or $u_k = \pm u$. It is clear that in these two cases one has $(au)u_k = (au_k)u$. Noting then that precisely six elements of the set B anticommute with u and that by (6), one has $\langle au, bu\rangle = \langle a, (bu)\bar{u}\rangle = \langle a, b|u|^2\rangle = \langle a, b\rangle$, we conclude. □

Suppose now as usual that the basis B coincides with the set $\{1, \mathbf{i}, \mathbf{j}, \mathbf{ij}, \mathbf{e}, \mathbf{ie}, \mathbf{je}, (\mathbf{ij})\mathbf{e}\}$, where $\mathbf{i} = u_1$, $\mathbf{j} = u_2$ and $\mathbf{e} = u_4$, so that for instance $u_5 = u_1 u_4$. Each element of the algebra $\mathbb{O}$ admits a unique expression as $q_1 + q_2\mathbf{e}$ with $q_1, q_2 \in \mathbb{H}$, where $\mathbb{H}$ is the quaternion algebra generated by $\mathbf{i}, \mathbf{j}$. Then the multiplication in $\mathbb{O}$ is defined by the standard multiplication relations in $\mathbb{H}$ and by the relations

$$q_1(q_2\mathbf{e}) = (q_2 q_1)\mathbf{e}, \quad (q_1\mathbf{e})q_2 = (q_1\bar{q}_2)\mathbf{e}, \quad (q_1\mathbf{e})(q_2\mathbf{e}) = -\bar{q}_2 q_1. \tag{9}$$

Put $B^0 = B\backslash\{u_0\}$. Let $\mathbf{i}', \mathbf{j}', \mathbf{e}'$ be three arbitrary distinct elements of the set $B^0 \cup (-B^0)$ such that $\mathbf{e}' \neq \pm\mathbf{i}'\mathbf{j}'$. Then there exists a unique automorphism Φ of the octonion algebra $\mathbb{O}$ such that $\Phi(\mathbf{i}') = u_1$, $\Phi(\mathbf{j}') = u_2$ and $\Phi(\mathbf{e}') = u_4$ (cf. [12, Lect. 15]). It is evident that $\Phi(u_0) = u_0$. Now, taking into account Lemmas 1 and 2 we obtain the relations (7).

Indeed, calculating $V(u_{\alpha_1}, u_{\alpha_2}, u_{\alpha_3}, u_{\alpha_4}; \beta, \gamma)$, by Lemma 2 we can suppose that $u_{\alpha_1} = u_0$. Since all elements $u_{\alpha_1} = u_0, u_{\alpha_2}, u_{\alpha_3}, u_{\alpha_4}$ are distinct, then according to either $u_{\alpha_4} = \pm u_{\alpha_2} u_{\alpha_3}$ or $u_{\alpha_4} \neq \pm u_{\alpha_2} u_{\alpha_3}$, we can obtain as image of the triple $u_{\alpha_2}, u_{\alpha_3}, u_{\alpha_4}$, under some automorphism Φ of $\mathbb{O}$, the triple u_1, u_2, u_3 or u_1, u_2, u_4, respectively.

To calculate $\tau(0, 1, 2, 3)$ consider the following 4-form (see definition (2) of $\tau_4(\omega)$):

$$\omega_{0123} \stackrel{\text{def}}{=} \omega_{01} \wedge \omega_{23} - \omega_{02} \wedge \omega_{13} + \omega_{03} \wedge \omega_{12}. \tag{10}$$

Denote by ω'_{ij} the restriction of the form ω_{ij} to the subspace $V \subset \mathbb{O}^2$ generated by the vectors X_k, for $k = 0, \dots, 7$. Let $\{x_0^*, \dots, x_7^*\}$ be the dual basis of V^*. Using the relations (9) it is easy to verify that

$$\begin{aligned}
\omega'_{01} &= x_0^* \wedge x_1^* + x_2^* \wedge x_3^* + x_4^* \wedge x_5^* - x_6^* \wedge x_7^*,\\
\omega'_{23} &= x_0^* \wedge x_1^* + x_2^* \wedge x_3^* - x_4^* \wedge x_5^* + x_6^* \wedge x_7^*,\\
\omega'_{02} &= x_0^* \wedge x_2^* - x_1^* \wedge x_3^* + x_4^* \wedge x_6^* + x_5^* \wedge x_7^*,\\
\omega'_{13} &= -x_0^* \wedge x_2^* + x_1^* \wedge x_3^* + x_4^* \wedge x_6^* + x_5^* \wedge x_7^*,\\
\omega'_{03} &= x_0^* \wedge x_3^* + x_1^* \wedge x_2^* + x_4^* \wedge x_7^* - x_5^* \wedge x_6^*,\\
\omega'_{12} &= x_0^* \wedge x_3^* + x_1^* \wedge x_2^* - x_4^* \wedge x_7^* + x_5^* \wedge x_6^*.
\end{aligned}$$

Calculations are very simple because for each item $\omega_{\beta_1\beta_2} \wedge \omega_{\beta_3\beta_4}$ in (10) the permutation $(\beta_1, \beta_2, \beta_3, \beta_4)$ of the set $\{0, 1, 2, 3\}$ satisfies the relation: $u_{\beta_1}u_{\beta_2} = \pm u_{\beta_3}u_{\beta_4}$. Now it is easy to verify that

$$\omega'_{01} \wedge \omega'_{23} = -\omega'_{02} \wedge \omega'_{13} = \omega'_{03} \wedge \omega'_{12} = 2x_0^* \wedge x_1^* \wedge x_2^* \wedge x_3^* + 2x_4^* \wedge x_5^* \wedge x_6^* \wedge x_7^*,$$

and, consequently,

$$\tau(0, 1, 2, 3) = (\omega_{0123} \wedge \omega_{0123})(X_0, \ldots, X_7) = 72.$$

To calculate $\tau(0, 1, 2, 4)$ consider the following 4-form (see definition (2) of $\tau_4(\omega)$):

$$\omega_{0124} \stackrel{\text{def}}{=} \omega_{01} \wedge \omega_{24} - \omega_{02} \wedge \omega_{14} + \omega_{04} \wedge \omega_{12}. \tag{11}$$

Since $\omega_{ij} = -\omega_{ji}$, we can rewrite the expression for $\omega_{0124} \wedge \omega_{0124}$ as

$$\begin{aligned}
&- 2\omega_{01} \wedge \omega_{02} \wedge \omega_{41} \wedge \omega_{42} - 2\omega_{01} \wedge \omega_{04} \wedge \omega_{21} \wedge \omega_{24} \qquad (12)\\
&- 2\omega_{02} \wedge \omega_{04} \wedge \omega_{12} \wedge \omega_{14} + (\omega_{01} \wedge \omega_{24})^2 + (\omega_{02} \wedge \omega_{14})^2 + (\omega_{04} \wedge \omega_{12})^2.
\end{aligned}$$

Remark that the first three terms of this sum are terms of the form $\omega_{ij} \wedge \omega_{ij'} \wedge \omega_{i'j} \wedge \omega_{i'j'}$ for some sequence of distinct elements $\{i, j, i', j'\} = \{0, 1, 2, 4\}$. Therefore, by [7, p.1170],

$$\begin{aligned}
-2\omega_{01} \wedge \omega_{02} \wedge \omega_{41} \wedge \omega_{42} - 2\omega_{01} \wedge \omega_{04} \wedge \omega_{21} \wedge \omega_{24}&\\
- 2\omega_{02} \wedge \omega_{04} \wedge \omega_{12} \wedge \omega_{14}(X_0, \ldots, X_7) &= 48.
\end{aligned}$$

Let us show that for each term above of the type $(\omega_{ij} \wedge \omega_{i'j'})^2$ we have $(\omega_{ij} \wedge \omega_{i'j'})^2(X_0, \ldots, X_7) = 8$.

Note that for any $i \neq j$ such that $u_ju_i \neq \pm u_0$ there exists some automorphism Φ of $\mathbb{O}$ such that $\Phi(\pm u_ju_i) = u_1$. Now taking into account the expression for the form ω'_{01} we obtain that

$$\omega'_{ij} = \varepsilon_0 x_{i_0}^* \wedge x_{i_1}^* + \varepsilon_2 x_{i_2}^* \wedge x_{i_3}^* + \varepsilon_4 x_{i_4}^* \wedge x_{i_5}^* + \varepsilon_6 x_{i_6}^* \wedge x_{i_7}^*,$$

where $\sigma_{ij} = (i_0, \ldots, i_7)$ is some permutation of the set $\{0, \ldots, 7\}$, $\varepsilon_{2k} = \pm 1$, and $\prod_{k=0}^{3} \varepsilon_{2k} \cdot \varepsilon(\sigma_{ij}) = -1$. Consider also the form

$$\omega'_{i'j'} = \varepsilon'_0 x^*_{j_0} \wedge x^*_{j_1} + \varepsilon'_2 x^*_{j_2} \wedge x^*_{j_3} + \varepsilon'_4 x^*_{j_4} \wedge x^*_{j_5} + \varepsilon'_6 x^*_{j_6} \wedge x^*_{j_7},$$

where $i \neq i'$ and $j \neq j'$.

We now prove two more lemmas. Remark that for the terms $(\omega_{ij} \wedge \omega_{i'j'})^2$ in (12), $u_i u_j \neq \pm u_{i'} u_{j'}$.

Lemma 3 *For arbitrary distinct elements $i, j, i', j' \in \{0, \ldots, 7\}$ such that $u_i u_j \neq \pm u_{i'} u_{j'}$, the 4-form $\omega'_{ij} \wedge \omega'_{i'j'}$ is a sum of at most eight linearly independent terms (4 – forms) $\omega'_{k,ij,i'j'}$, $k = 0, \ldots, 7$, of type $\pm x^*_{k_0} \wedge x^*_{k_1} \wedge x^*_{k_2} \wedge x^*_{k_3}$. For each such term $\omega'_{k,ij,i'j'}$, there is a unique term $\varepsilon_{2p} x^*_{i_{2p}} \wedge x^*_{i_{2p+1}}$ of ω'_{ij} and a unique term $\varepsilon'_{2p'} x^*_{j_{2p'}} \wedge x^*_{j_{2p'+1}}$ of $\omega'_{i'j'}$ such that their exterior product is proportional to $\omega'_{k,ij,i'j'}$ (and, consequently, it is equal to $\omega'_{k,ij,i'j'}$).*

Proof Put $u_l = \pm u_i u_j$ and $u_{l'} = \pm u_{i'} u_{j'}$. By the assumptions of the lemma u_l and $u_{l'}$ are two distinct imaginary units of $\mathbb{O}$. Therefore if $\omega'_{ij}(u_{i_0}, u_{i_1}) = \pm \langle u_{i_0}, u_l u_{i_1} \rangle \neq 0$ then $u_l = \pm u_{i_0} u_{i_1}$ and $u_{l'} \neq \pm u_{i_0} u_{i_1}$, i.e. $\omega'_{i'j'}(u_{i_0}, u_{i_1}) = 0$. So precisely two terms of $\omega'_{i'j'}$ contain $x^*_{i_0}$ and $x^*_{i_1}$ as a factor. Therefore there exists precisely two terms of $\omega'_{i'j'}$ such that their exterior product with $x^*_{i_0} \wedge x^*_{i_1}$ is not zero. Since the form ω'_{ij} contains four terms, the number of linearly independent terms of $\omega'_{ij} \wedge \omega'_{i'j'}$ is at most eight.

Assume that the product of the terms $x^*_{i_0} \wedge x^*_{i_1}$ and $x^*_{j_0} \wedge x^*_{j_1}$ of the forms ω'_{ij} and $\omega'_{i'j'}$ respectively, is not trivial, i.e. $\{i_0, i_1\} \cap \{j_0, j_1\} = \emptyset$. The forms ω'_{ij} and $\omega'_{i'j'}$ contain a unique term with the factor $x^*_{i_0}$. As we show above, in the form $\omega'_{i'j'}$ the second factor of this term is not equal to $x^*_{i_1}$. Assume that this factor is equal to $x^*_{j_k}$, $k = 0, 1$. Then $\omega'_{i'j'}(u_{i_0}, u_{j_k}) \neq 0$, i.e. $u_{i_0} = \pm u_{l'} u_{j_k}$. But $u_{j_0} = \pm u_{l'} u_{j_1}$, i.e. $\{i_0, i_1\} \cap \{j_0, j_1\} \neq \emptyset$. This contradicts our non-triviality assumption. We can proceed similarly in the case of the factor $x^*_{i_1}$. □

Lemma 4 *For any distinct elements $i, j, i', j' \in \{0, \ldots, 7\}$ such that $u_i u_j \neq \pm u_{i'} u_{j'}$, the expression $V(u_i, u_j, u_{i'}, u_{j'}) = 2^{-4} \sum_{\sigma \in S_8} C_\sigma(u_i, u_j, u_{i'}, u_{j'})$, contains at most $2^4 \cdot 8$ non-zero terms.*

Proof By the previous lemma, each term of $\omega'_{ij} \wedge \omega'_{i'j'}$ is the exterior product of a uniquely defined pair of terms of the forms ω'_{ij} and $\omega'_{i'j'}$. On the other hand, this term of $\omega'_{ij} \wedge \omega'_{i'j'}$ determines a unique complementary factor in $x^*_0 \wedge \cdots \wedge x^*_7$ which belongs to $\omega'_{ij} \wedge \omega'_{i'j'}$. If such a factor exists, by the previous lemma this factor is the exterior product of a uniquely defined pair of terms of the forms ω'_{ij} and $\omega'_{i'j'}$. Since the number of terms of $\omega'_{ij} \wedge \omega'_{i'j'}$ equals at most 8 and due to the skew-symmetry of the 2-forms, the lemma follows. □

Suppose that $i, j, i', j' \in \{0, \ldots, 7\}$ are distinct elements such that $u_i u_j \neq \pm u_{i'} u_{j'}$. Due to the skew-symmetry of the 2-forms, one has $V(u_i, u_j, u_{i'}, u_{j'}) = \sum_{[\sigma]\in S_8'} C_\sigma(u_i, u_j, u_{i'}, u_{j'})$, where $S_8' = S_8/S'$ and the subgroup $S' \subset S_8$ is generated by the 4 transpositions $(0,1)$, $(2,3)$, $(4,5)$, and $(6,7)$. By Lemma 4 this sum contains at most 8 non-zero terms. Let us describe these terms. To this end, using (5) we can rewrite the expression for $C_\sigma(u_i, u_j, u_{i'}, u_{j'})$ as

$$-\varepsilon(\sigma)\langle u_{i_0}u_i, u_{i_1}u_j\rangle\langle u_{i_2}u_{i'}, u_{i_3}u_{j'}\rangle\langle u_{i_4}u_i, u_{i_5}u_j\rangle\langle u_{i_6}u_{i'}, u_{i_7}u_{j'}\rangle,$$

since $\bar{u}_k = -u_k$ for all of the seven imaginary units. Let $u \in B$ and $a \in B^{\pm}$. Arguing as in the proof of Lemma 2, we obtain that if $au = -ua$ then $(u_k a)u = (-u_k u)a$. But $au \neq -ua$ if and only if $a = \pm u$ or $a = \pm u_0$ or $u = \pm u_0$. In all these cases $(u_k a)u = (u_k u)a$. Since $\langle au, bu\rangle = \langle a, b\rangle$, we obtain the following expression for $C_\sigma(u_{i_0}, u_i, u_{i_1}, u_j)$:

$$-\varepsilon(\sigma)\big\langle(u_{i_0}u)u_i, (u_{i_1}u)u_j\big\rangle\big\langle(u_{i_2}u)u_{i'}, (u_{i_3}u)u_{j'}\big\rangle \cdot \big\langle(u_{i_4}u)u_{i}, (u_{i_5}u)u_j\big\rangle\big\langle(u_{i_6}u)u_{i'}, (u_{i_7}u)u_{j'}'\big\rangle$$

(the elements $u_{i_0}u_i$, $u_{i_1}u_j$ occur in this expression twice).

Suppose now that $C_\sigma(u_i, u_j, u_{i'}, u_{j'}) \neq 0$ for some $\sigma \in S_8$. Right multiplication by u determines the permutation σ^u of the set B: $u_k u = \varepsilon^u_{u_k}\sigma^u(u_k)$ ($\varepsilon^u_{u_k} = \pm 1$). This permutation is even since if $u \neq u_0$ then $u^2 = -u_0$ and σ^u is a product of four independent transpositions. The sequence $(\varepsilon^u_{u_0}, \ldots, \varepsilon^u_{u_7})$ contains an even number of -1 (see [7, p. 1170]).

Thus $C_\sigma(u_i, u_j, u_{i'}, u_{j'}) = C_{\sigma^{u_k}\sigma}(u_i, u_j, u_{i'}, u_{j'})$ for all of the eight even permutations σ^{u_k}, $k = 0, \ldots, 7$. It only remains to be proved that the permutations $\sigma^{u_k}\sigma$ determine distinct classes in the quotient group S_8'.

Suppose that $\sigma^{u_k}\sigma = \sigma^{u_p}\sigma \cdot s$ for some element $s \in S_8'$ and $k \neq p$. Taking into account that $\sigma^{u_p}\sigma^{u_k} = \sigma^{u_k}\sigma^{u_p} = \sigma^{u_q}$, where $u_q \in B$ and $u_q = \pm u_k u_p = \pm u_p u_k$, we can assume that $u_p = u_0$ and $\sigma(u_0) = u_0$. But for $u \in B$ we have $\{\pm u_0 u, \pm u_{i_1} u\} = \{\pm u_0, \pm u_{i_1}\}$ if and only if $u \in \{u_0, u_{i_1}\}$. Since $C_\sigma(u_i, u_j, u_{i'}, u_{j'}) \neq 0$, we have $u_{i_1} = u_l$ and $u_{i_3} = \pm u_{l'} u_{i_2}$, where $u_l = \pm u_i u_j$ and $u_{l'} = \pm u_{i'} u_{j'}$. Taking into account that $u_l \neq u_{l'}$, we obtain that $u_{i_3} \neq \pm u_l u_{i_2} = \pm u_{i_1} u_{i_2}$, i.e. $u_k = u_0$, a contradiction.

So $V(u_i, u_j, u_{i'}, u_{j'}) = 8C_\sigma(u_i, u_j, u_{i'}, u_{j'})$, where $\sigma \in S_8$ is an arbitrary permutation such that $C_\sigma(u_i, u_j, u_{i'}, u_{j'}) \neq 0$. Using now the relations (9), we can show that there exists a common odd permutation $\sigma = (0, 1, 2, 4, 3, 5, 6, 7)$ for the following sequences (i, j, i', j'): $(0, 1, 2, 4)$, $(0, 2, 1, 4)$ $(0, 4, 1, 2)$. For all these cases $C_\sigma(u_i, u_j, u_{i'}, u_{j'}) = 1$ and, consequently,

$$(\omega_{01} \wedge \omega_{24})^2 + (\omega_{02} \wedge \omega_{14})^2 + (\omega_{04} \wedge \omega_{12})^2(X_0, \ldots, X_7) = 24.$$

Thus $\tau(0, 1, 2, 4) = (\omega_{0124} \wedge \omega_{0124})(X_0, \dots, X_7) = 72$ and by expressions (2) and (7), one has

$$\tau_4(\omega)(X_0, \dots, X_7) = \binom{8}{4} \cdot 72 = 70 \cdot 72 = 5040,$$

so concluding the proof of Proposition 1.

Noting now that $\Omega_0^8(X_0, \dots, X_7) = -14 \cdot 1440 = -4 \cdot 5040$, Corollary 1 follows.

References

1. Abe, K.: Closed regular curves and the fundamental form on the projective spaces. P. Jpn. Acad. A-Math. **68**(6), 123–125 (1992)
2. Abe, K., Matsubara, M.: Invariant forms of the exceptional symmetric spaces *FII* and *EIII*. Transformation Group Theory, pp. 3–16. Korea Adv. Inst. Sci. Tech, Taejŏn (1996)
3. Alekseevskiĭ, D.V.: Riemannian spaces with non-standard holonomy groups. Funct. Anal. Appl. **2**, 97–105 (1968)
4. Brada, C., Pécaut-Tison, F.: Géométrie du plan projectif des octaves de Cayley. Geom. Dedic. **23**(2), 131–154 (1987)
5. Brada, C., Pécaut-Tison, F.: Calcul explicite de la courbure et de la 8-forme canonique du plan projectif des octaves de Cayley. C. R. Acad. Sci. A Math. **301**(2), 41–44 (1985)
6. Brown, R.B., Gray, A.: Riemannian manifolds with holonomy group Spin(9). In: Diff. Geom. honor of K. Yano, pp. 41–59. Kinokuniya, Tokyo (1972)
7. Castrillón López, M., Gadea, P.M., Mykytyuk, I.V.: The canonical eight-form on manifolds with holonomy group Spin(9). Int. J. Geom. Methods M. **7**(7), 1–25 (2010)
8. Friedrich, Th: Weak Spin(9)-structures on 16-dimensional riemannian manifolds. Asian J. Math. **5**(1), 129–160 (2001)
9. Friedrich, Th: Spin(9)-structures and connections with totally skew-symmetric torsion. J. Geom. Phys. **47**(2–3), 197–206 (2003)
10. Lam, K.-H.: Spectrum of the Laplacian on manifolds with Spin(9) holonomy. Math. Res. Lett. **15**(5–6), 1167–1186 (2008)
11. Parton, M., Piccinni, P.: Spin(9) and almost complex structures on 16-dimensional manifolds. Ann. Glob. Anal. Geom. **41**(3), 321–345 (2012)
12. Postnikov, M.: Lie groups and Lie algebras. Lectures in Geometry. Semester V. Translated from the Russian by Vladimir Shokurov. MIR, Moscow (1986)

Flaws in the Application of Number Theory in Key Distribution Schemes for Multicast Networks

A. Peinado

Dedicated to Jaime Muñoz Masqué, unquestionable reference and irreplaceable guide in science and life on the occassion of his 65th birthday

Abstract In this note, an interesting trend about the way in which the number theory in multicast networks is often applied, is reported. Surprisingly, in recent years, some new proposals for key distribution schemes are still proposed employing very similar erroneous concepts than those applied in 1999, which were already reported by professor Muñoz-Masqué in 2005. Some apparently well-constructed cryptographic equations suffer from a real weakness due to a flaw in the definition of the cryptographic keys, allowing to perform an easy factorization and, as a consequence, the recovering of the user's keys. Thirteen years later, very similar weaknesses arise.

Keywords Integer factorization · Cryptanalysis · Key distribution scheme · Multicast

1 Introduction

Unicast communication stands for the traditional scheme in which the information is transmitted from one host to another. It is known as *one-to-one* communication. However, when the same information has to be transmitted to many users, the unicast architecture turns inefficient because the transmitter sends multiples copies of the same information. This is the case of live TV channels on Internet where many users are watching the same content simultaneously. This causes the increasing of data

A. Peinado (✉)
Departamento Ingeniería de Comunicaciones, Andalucía Tech.,
Universidad de Málaga, 29071 Málaga, Spain
e-mail: apeinado@ic.uma.es

M. Castrillón López et al. (eds.), *Geometry, Algebra and Applications: From Mechanics to Cryptography*, Springer Proceedings in Mathematics & Statistics 161, DOI 10.1007/978-3-319-32085-4_16

traffic that could congest the entire network when the number of target users is high. Furthermore, a computational overhead appears in the transmitter if the information is encrypted, because each data frame or packet will be encrypted with a different key in order to be decrypted by each user.

The alternative approach is the Multicast communication, defined as the transmission of information from one host to multiple hosts, and known as *one-to-many* communication. The advantages over the unicast communication turn more evident in cases of huge numbers of receivers, such as multimedia communications. Multicast operation sends only one message that will be duplicated following a tree-based structure in order to reach the target users using the least numbers of messages. The information is duplicated only when a bifurcation appears in the path to the multiple destinations.

In general, secure multicast requires specific protocols and schemes, different from those applied in classical unicast communications. One of the most important and complex issues in secure multicast is the key management because of the network topology and the amount of users that join or leave the multicast group continuously. The encryption keys must be periodically renewed to avoid external attacks and to provide forward and backward secrecy, so as to the users who joins to the system cannot decrypt previous contents and the users who leaves the system cannot decrypt future contents.

Many proposals for multicast/broadcast encryption and key distribution have been presented from the last century, with the following objectives in mind: minimization of the number of messages between the hosts; reducing the size of cryptographic parameters and the messages; increasing of efficiency in cryptographic operations in order to reduce the computation time; simplification of the processes related to the joining and/or leaving of users; and providing forward and backward secrecy.

One of those proposals is the one presented by Liaw in 1999 [3]. It is a centralized system with a Central Authority Server (CAS) which generates the keys of each user that participates in the multicast scheme. That system, based on a multiplicative group key, reduces the number of messages, thus facilitating the joining operation. However, Sun [10] proved that it cannot be operated because a very large amount of information (2^{71} bits) must be kept by each user and be sent for each broadcast. Later, Tseng and Jan [11] founded several weaknesses on the Liaw's cryptosystem and proposed a modification in 2001. Muñoz-Masqué and Peinado [5] reported in 2005 an inconsistency in the improvement of Tseng and Jan, provided a new cryptanalysis of both the original and the improvement scheme, and presented a new modification that overcomes the previous known attacks.

Many years later, in 2012, Naranjo et al. [6] proposed a centralized key distribution scheme based on the extended Euclidean algorithm [2]. The main advantage of this scheme is very similar to that of mentioned protocols, that is, only one message is generated at each rekeying operation and only one long-term key is associated to each user. This similarity has produced similar flaws, reported by Peinado and Ortiz [7] in 2013. Despite of that, Vijayakumar et al. [12] has proposed again a slight modification of this algorithm using the same approach and suffering from the same flaws.

Next section describes the Liaw's cryptosystem and the improvement proposed by Tseng and Jan. Section 3 shows the flaws detected in the application of the number theory. Section 4 deals with the recent secure multicast proposals, their similarities with the previous schemes and the flaws that once more are presented in this kind of schemes. Finally, Sect. 5 shows the conclusions.

2 Liaw's Cryptosystem and Its Improvement

In this section, the Liaw's original cryptosystem and its modification are described, using the same notations as in [3, 11]. In both cases, the protocol is composed by three phases, in which a CAS and n users U_i, $1 \leq i \leq n$, interact.

2.1 Liaw's Cryptosystem

System setup phase. This is a previous phase to generate the necessary parameters. The CAS generates the private and public keys of the system that allow the users to communicate with the CAS. This pair of keys is generated using an RSA scheme [9]. Hence, the CAS computes the modulus $N = p \cdot q$, where $p = 2p' + 1, q = 2q' + 1$, are safe prime numbers, i.e., p, q, p', q' are all prime. We set $\lambda(N) = \text{lcm}(p - 1, q - 1)$ and denote the Euler totient function by $\phi(N)$. The integers d, e are selected such that $d \cdot e \equiv 1 \pmod{\phi(\lambda(N))}$. Hence, N and d are made public, whereas p, q and e are kept secret.

Next, the CAS generates a pair of keys for each user U_i. A secret integer K_0 is chosen to compute the private key (t_i, K_i) and the public key $f(t_i)$ such that

$$\begin{aligned} K_i &= K_0^{t_i} \pmod{N} \\ f(t_i) &= t_i^e \end{aligned} \tag{1}$$

where t_i is prime.

Broadcasting phase. Without loss of generality, let us suppose than U_1 is the user who wants to transmit data to the group of users $U_2, U_3, \ldots, U_a$. Then, U_1 sends a request to the CAS in order to generate the encryption key MK_1. To do this, the CAS computes the following parameters:

$$\begin{aligned} f(B_1) &= B_1^e \\ MK_1 &= K_0^{B_1} \pmod{N} \\ PK_1 &= E_{t_1}(MK_1) \end{aligned} \tag{2}$$

where $B_1 = t_2 \cdot t_3 \cdots t_a$ and $E_k(.)$ is the symmetric encryption function of the system with key k. Next, CAS sends $f(B_1)$ and PK_1 to the user U_1 and $f(B_1)$ to every user $U_i, 2 \leq i \leq a$. When U_1 receives PK_1, he can recover the encryption key as

$$MK_1 = D_{t_1}(PK_1). \tag{3}$$

The user U_1 encrypts the message M as $C = E_{MK_1}(M)$. Finally, he broadcasts C.

Decryption phase. When a user $U_j, 2 \leq j \leq a$, receives $f(B_1)$ and C, the secret key MK_1 must be obtained to decrypt C. Hence, U_j performs the following computation:

$$\begin{aligned} MK_1 &= K_j^{(f(B_1)/f(t_j))^d} \pmod N = K_0^{t_j \cdot \left(\prod_{i \neq j, 2 \leq i \leq a} t_i^e \right)^d} \pmod N \\ &= K_0^{t_2 \cdot t_3 \cdots t_a} \pmod N = K_0^{B_1} \pmod N . \end{aligned} \tag{4}$$

2.2 *Improvement of Liaw's Cryptosystem*

Only two modifications were proposed by Tseng and Jan [11] to improve the original system. On the one hand, the private key t_i is only known to the CAS. Therefore, the private key and public key of every user U_i are now K_i and $f(t_i)$, respectively. This modification tries to avoid a conspiracy attack to obtain K_0. Note that t_i is no longer known by the user U_i.

On the other hand, the function f is redefined as $f(x) = x^e \pmod{\lambda(N)}$. Hence, the public key $f(t_i)$ of user U_i and $f(B_1)$ are computed as

$$\begin{aligned} f(t_i) &= t_i^e \pmod{\lambda(N)}, \\ f(B_1) &= (t_2 \cdot t_3 \cdots t_a)^e \pmod{\lambda(N)}. \end{aligned} \tag{5}$$

3 Flaws in the Liaw-Type Cryptosystems

In this section, the different flaws detected in the previous systems by Tseng and Jan [11], Sun [10] and Muñoz-Masqué and Peinado [5] are presented. All of them are related with misapplications of the number theory.

Flaw 1. (*Detected in Liaw's cryptosystem* [3]. *Reported by Tseng and Jan* [11], *Sun* [10] *and Muñoz-Masqué and Peinado* [5]) In [3] the parameter $f(t_i) = t_i^e$ is a component of the public key of the user U_i. Although it looks like an RSA system, one can observe that no modular operation is applied. Hence, publishing the value $f(t_i) = t_i^e$ compromises the security, though factoring integers is a hard problem, detecting whether a given integer is a prime power is not that hard. In fact, if $k = \pi^e$, π being a prime, then by Fermat's theorem we have

$$b^k = (F \circ \overset{e}{\ldots} \circ F)(b) \equiv b \pmod{\pi} \tag{6}$$

for every integer b, where F is the function defined by $F(x) = x^\pi$. Hence, π divides $\gcd(b^k - b, k)$ and for most values of b we will even have $\gcd(b^k - b, k) = \pi$. The running time for the Euclid algorithm is $O((\ln k)^2)$ and computing $b^k \pmod{k}$ is $O((\ln k)^2 \ln b)$. Hence, $k = \pi^e$ can be factored in $O((\ln k)^3)$ (see [1], Algorithm 1.7.4).

Flaw 2. (*Detected in Tseng's cryptosystem* [11]. *Reported by Muñoz-Masqué and Peinado* [5]) Tseng and Jan proposed the utilization of the value $f(t_i) = t_i^e \pmod{\lambda(N)}$ instead of $f(t_i) = t_i^e$, in order to increase the security level. However, the procedure to recover the encryption key (Eq. 4) suffers from an inconsistency. A simple numerical example shows that the quotient $f(B_1)/f(t_i)$ may not be an integer and hence the decryption process could fail. More precisely, the procedure to recover the key MK_1 is not valid as the quotient $f(B_1)/f(t_i)$ does not always exist in $\mathbb{Z}$.

4 Recent Proposals

Naranjo et al. [6], in 2012, and Vijayakumar et al. [12], in 2013, proposed very similar centralized key distribution schemes based on the extended Euclidean algorithm for multicast communications. In both schemes, the CAS generates a multicast encryption key, that is distributed to the users. To do so, the CAS generates a long-term secret key K_i for each user U_i. The main parameter of this scheme is L defined as

$$L = \prod_{i=1}^{n} K_i \tag{7}$$

where L is not a public parameter; it is a secret value of the CAS.

4.1 Similarities with Liaw-Type Cryptosystems

Although the protocols in [6, 12] are different to Liaw's cryptosystem, there exist several similarities that determine similar effects when they are analyzed from a cryptographic point of view. The analogy can be summarized in two main items.

- No modular operation is performed to compute the main parameters. This affects to the size of the parameters, increasing dramatically. Furthermore, it is a source of insecurity as it is reported in the next subsection.
- The participation of the users is multiplicative; that is, the procedure defined to recover the encryption key requires to perform a multiplication of some parameter of each user. In Liaw's cryptosystem $B_1 = t_2 \cdot t_3 \cdots t_a$, and in the recent proposals L is the product of all the secret keys (Eq. 7).

4.2 Flaw of Recent Proposals

On the one hand, the secret long-term key of the users is defined to be 64-bit long; a prime number in the case of [6], and an integer co-prime with the other keys, in the case of [12]. In any case, this bit length allows a brute force attack. Probably, the keys are not larger in order to bound the size of the parameters computed from the product of all keys.

On the other hand, as it is reported in [7], the secret parameter L can be obtained in most cases, but the most important weakness is that a multiple of L can always be obtained. As a consequence, the factorization of L could be performed to get the user's key [4].

The flaw is a combination of a low key size and the multiplicative operation without modular reduction. Although the size of L is very large, (6400 bits long for 64-bit keys and a hundred of users, see [6]), it does not provide a real protection. It is important to note that the factorization problem is reduced to find prime factors of 64 bits. Hence, it is not a general factorization problem. In [7], a genetic algorithm is applied to find those prime factors. The case of [12] if trivial, since small factors are present in L. The Pollard's rho method [8] can be applied to recover them.

5 Conclusion

The big numbers often convey a false sense of security. For that reason, we can observe how the cryptographic protocols and schemes keep proposing the utilization of big numbers as a method to provide security and protection. Sometimes, the specific operational conditions are the source of this erroneous design. Multicast communication is a representative example where a very big numbers are generated (the dummy security parameter) from the product of many small numbers (the secret key of users). This approach is employed due to the limitation in size of message and parameters that a user has to store and transmit. In most cases, the definition of the computations in a finite field reduces the size of the parameters and increases the security. This note is an evidence that big numbers are not always secure.

Acknowledgments This work has been partially supported by project TIN2014-55325-C2-1-R (ProCriCiS), funded by Ministerio de Economía y Competitividad, Spain.

References

1. Cohen, H.: A Course in Computational Algebraic Number Theory. Graduate Text in Maths 138. Springer, Berlin (1993)
2. Gathen, J., Gerhard, J.: Modern Computer Algebra. Cambridge University Press, Cambridge (1999)

3. Liaw, H.T.: Broadcasting cryptosystem in computer networks. Comput. Math. Appl. **37**, 85–87 (1999)
4. Menezes, A., Oorschot, P., Vanstone, S.A.: Handbook of Applied Cryptography. CRC Press, Boca Raton (1996)
5. Muñoz Masqué, J., Peinado, A.: Cryptanalysis of improved Liaw's broadcasting cryptosystem. J. Inf. Sci. Eng. **22**, 391–399 (2006)
6. Naranjo, J.A.M., Antequera, N., Casado, L.G., López-Ramos, J.A.: A suite of algorithms for key distribution and authentication in centralized secure multicast environments. J. Comput. Appl. Math. **236**, 3042–3051 (2012)
7. Peinado, A., Ortiz, A.: Cryptanalysis of a key refreshment scheme for multicast protocols by means of genetic algorithm. Log. J. IGPL **21**(4), 671–679 (2013)
8. Pollard, J.M.: A Monte Carlo method for factorization. BIT Numer. Math. **15**(3), 331–334 (1975)
9. Rivest, R., Shamir, A., Adleman, L.: A method for obtaining digital signatures and public-key cryptosystems. Commun. ACM **21**, 120–126 (1978)
10. Sun, H.M.: Security of broadcasting cryptosystem in computer networks. Electron. Lett. **35**, 2108–2109 (1999)
11. Tseng, Y.M., Jan, J.K.: Cryptanalysis of Liaw's broadcasting cryptosystem. Comput. Math. Appl. **41**, 1575–1578 (2001)
12. Vijayakumar, P., Bose, S., Kannan, A.: Centralized key distribution protocol using the greatest common divisor method. Comput. Math. Appl. **65**, 1360–1368 (2013)

Einstein–Hilbert Lagrangian Induced on the Linear Frame Bundle

Luis Pozo and Eugenia Rosado

Dedicated to Jaime Muñoz Masqué, from our deepest gratitude and friendship, on the occasion of his 65th birthday.

Abstract Let $p_F: FN \to N$ be the bundle of linear frames of a C^∞ manifold N. The Lagrangian induced on FN by the Einstein–Hilbert Lagrangian is written as a differentiable function of the system of Diff$N$-invariant Lagrangian defined on the linear frame bundle.

Keywords Bundle of metrics · Bundle of linear frames · Einstein-Hilbert Lagrangian · Linear frame bundle · Jet bundles

1 Introduction

Let $p_{\mathcal{M}}: \mathcal{M} = \mathcal{M}(N) \to N$ be the bundle of pseudo-Riemannian metrics of a given signature (n^+, n^-), $n^+ + n^- = n = \dim N$, over a connected C^∞ manifold N oriented by a volume form $\mathbf{v} \in \Omega^n(N)$. The Einstein–Hilbert functional is the second-order Lagrangian density $L_{EH}\mathbf{v}$ on $\mathcal{M}$ defined along a metric g by $S^g \mathbf{v}_g$, where S^g denotes the scalar curvature of g and $\mathbf{v}_g$ its Riemannian volume form; namely,

L. Pozo
Departamento de Matemática Aplicada a las TIC, Escuela Técnica Superior de Ingeniería de Sistemas Informáticos, UPM, Ctra. Valencia, Km. 7, 28031 Madrid, Spain
e-mail: lm.pozo@upm.es

E. Rosado (✉)
Departamento de Matemática Aplicada, Escuela Técnica Superior de Arquitectura, UPM, Avda. Juan de Herrera 4, 28040 Madrid, Spain
e-mail: eugenia.rosado@upm.es

M. Castrillón López et al. (eds.), *Geometry, Algebra and Applications: From Mechanics to Cryptography*, Springer Proceedings in Mathematics & Statistics 161, DOI 10.1007/978-3-319-32085-4_17

$$L_{EH} \circ j^2 g = \sqrt{|\det(g_{ab})|} g^{jk} \left\{ \frac{\partial (\Gamma^g)^i_{jk}}{\partial x^i} - \frac{\partial (\Gamma^g)^i_{ik}}{\partial x^j} + (\Gamma^g)^l_{jk} (\Gamma^g)^i_{il} - (\Gamma^g)^l_{ik} (\Gamma^g)^i_{jl} \right\},$$

where we confine ourselves to consider coordinate systems $(x^1, \dots, x^n)$ on M adapted to $\mathbf{v}$, i.e.,

$$\mathbf{v} = dx^1 \wedge \cdots \wedge dx^n, \quad \mathbf{v}_g = \sqrt{|\det(g_{ab})|}\mathbf{v}, \quad g = g_{ab} dx^a \otimes dx^b,$$

and where $(\Gamma^g)^i_{jk}$ are the Christoffel symbols of the Levi-Civita connection ∇^g of the metric g.

The bundle of metrics $p_{\mathcal{M}} : \mathcal{M} \to N$ can be viewed as the quotient bundle FN/G, where FN is the bundle of linear frames of N and $G = O(n^+, n^-) \subset GL(n; \mathbb{R})$ is the orthogonal group corresponding to the signature (n^+, n^-). Hence every Lagrangian L on $J^r(\mathcal{M})$ induces a Lagrangian $\bar{L}$ on $J^r(FN)$ in a natural way, which is Diff$N$-invariant if L is Diff$N$-invariant. Accordingly, the problem of determining Diff$N$-invariant Lagrangians on the bundle of metrics reduces to that of determining Diff$N$-invariant Lagrangians on the bundle of linear frames which, in addition, are projectable onto the quotient bundle $q : FN \to FN/G \cong \mathcal{M}$. Let $\mathcal{D}^r$ be the involutive distribution on $J^r(FN)$ spanned by the natural lifts of vector fields on N, and let $\mathcal{V}^r = \ker J^r(q)_*$ be the involutive distribution of $J^r(q)$-vertical vector fields. A Lagrangian $\bar{L}$ on $J^r(FN)$ is Diff$N$-invariant if it is a first integral of the distribution $\mathcal{D}^r$, and it is projectable onto $J^r(\mathcal{M})$ if it is a first integral of $\mathcal{V}^r$. Hence, Diff$N$-invariant Lagrangians on $J^r(\mathcal{M})$ can be identified to the first integrals of the involutive distribution $\mathcal{D}^r + \mathcal{V}^r$. Moreover, $\mathcal{V}^r$ has a simple gauge interpretation: $\mathcal{V}^r_{j^r_x s} \cong J^r_x(N, \mathfrak{g})$, where $\mathfrak{g}$ is the Lie algebra of G, thus providing a similar meaning for the r-jet prolongation as the usual one in gauging $\mathfrak{g}$. Essentially, this is the geometry behind the theory that considers General Relativity as a gauge theory from the initial works by Kibble, Sciama, etc. (e.g., see [3, 8]); i.e., one first obtains Diff$N$-invariant Lagrangians on the bundle of linear frames and then, one imposes them to be invariant under the "gauge algebra" $\mathcal{V}^r$. Of course, one can work directly with orthonormal frames, specially in the $1+3$ approach (e.g., see [9]) but the former setting seems to be very suitable in dealing with a spacetime N with no preferred geometric decomposition. In any case, such an approach has the advantage of separating diffeomorphism invariance—a purely geometric condition—from the invariance under the group G.

In [1], P.L. García and J. Muñoz Masqué determined a canonical basis for the rings of r-order differential invariants of linear frames on a differentiable manifold N with respect to the Lie algebra of the vector fields of N.

The purpose of the present paper is to write the Lagrangian induced on $J^2(FN)$ by the Einstein–Hilbert Lagrangian as a differentiable function of the basis of Diff$N$-invariant Lagrangians defined on $J^2(FN)$.

2 Preliminaries and Notations

2.1 Jet-Bundle Notations

Let $p\colon E \to N$ be an arbitrary fibred manifold and let $p^k\colon J^kE \to N$ be the bundle of k-jets of local sections of p, with projections $p^k_l\colon J^kE \to J^lE$, $k \geq l$. Let $m = \dim E - \dim N$. Every fibred coordinate system (x^j, y^α), $1 \leq j \leq n$, $1 \leq \alpha \leq m$, for p induces a coordinate system (x^j, y^α_I), on the r-jet bundle, where $I = (i_1, \ldots, i_n) \in \mathbb{N}^n$ is an integer multi-index of order $|I| = i_1 + \cdots + i_n \leq r$, given by,

$$y^\alpha_I\left(j^r_x s\right) = \frac{\partial^{|I|}(y^\alpha \circ s)}{\partial(x^1)^{i_1}\ldots\partial(x^n)^{i_n}}(x),$$

where s is a local section of p defined on a neighbourhood of $x \in N$. We set $(j) = (0, \ldots, 0, \overset{(j)}{1}, 0, \ldots, 0) \in \mathbb{N}^n$, $(jk) = (j) + (k)$, etc., and $y^\alpha_0 = y^\alpha$.

Every morphism $\Phi\colon E \to E'$ whose associated map $\phi\colon N \to N'$ is a diffeomorphism, induces a map

$$\begin{aligned} &\Phi^{(r)}\colon J^rE \to J^rE', \\ &\Phi^{(r)}(j^r_x s) = j^r_{\phi(x)}(\Phi \circ s \circ \phi^{-1}). \end{aligned} \tag{1}$$

If Φ_t is the flow of a vector field $X \in \mathrm{aut}(p)$, then $\Phi^{(r)}_t$ is the flow of a vector field $X^{(r)} \in \mathfrak{X}(J^rE)$, called the infinitesimal contact transformation of order r associated to the vector field X. The mapping $\mathrm{aut}(p) \ni X \mapsto X^{(r)} \in \mathfrak{X}(J^rE)$ is an injection of Lie algebras.

2.2 The Bundle of Linear Frames

Let $p_F\colon FN \to N$ be the bundle of linear frames of N. Each coordinate system (x^i) on an open domain $U \subseteq N$ induces a coordinate system (x^i, x^i_j) on $(p_F)^{-1}(U)$, where the functions x^i_j are defined by,

$$u = \left((\partial/\partial x^1)_x, \ldots, (\partial/\partial x^n)_x\right) \cdot \left(x^i_j(u)\right), \; x = p_F(u), \forall u \in (p_F)^{-1}(U),$$

or equivalently,

$$u = (X_1, \ldots, X_n) \in F_x(N), \; X_j = x^i_j(u)\left(\frac{\partial}{\partial x^i}\right)_x, \quad 1 \leq j \leq n,$$

and also a coordinate system $(x^i, x^i_j, x^i_{j,(k_1\dots k_q)})$, $1 \le q \le r$, $k_1 \le \dots \le k_q$, on $J^r(FU)$ defined by,

$$x^i_{j,(k_1\dots k_q)}(j^r_x s) = \frac{\partial^q (x^i_j \circ s)}{\partial x^{k_1}\dots\partial x^{k_q}}(x).$$

2.3 *The Bundle of Metrics*

Let $p_{\mathscr{M}} \colon \mathscr{M} = \mathscr{M}(N) \to N$ be the bundle of pseudo-Riemannian metrics of a given signature (n^+, n^-), $n^+ + n^- = n$ on N. Every coordinate system (x^i) on an open domain $U \subseteq N$ induces a coordinate system (x^i, y_{jk}) on $(p_{\mathscr{M}})^{-1}(U)$, where the functions $y_{jk} = y_{kj}$ are defined by,

$$g_x = \sum_{i\le j} y_{ij}(g_x)(dx^i)_x \otimes (dx^j)_x, \ \forall g_x \in (p_{\mathscr{M}})^{-1}(U), \tag{2}$$

and also a coordinate system $(x^i, y_{ij}, y_{ij,(k_1\dots k_q)})$, $1 \le q \le r$, $k_1 \le \dots \le k_q$, on $J^r((p_{\mathscr{M}})^{-1}(U))$ defined by,

$$y_{ij,(k_1\dots k_q)}(j^r_x s) = \frac{\partial^q (y_{ij} \circ s)}{\partial x^{k_1}\dots\partial x^{k_q}}(x).$$

2.4 *Natural Lifts*

Let $f_{\mathscr{M}} \colon \mathscr{M} \to \mathscr{M}$, cf. [7] (resp. $\tilde{f} \colon FN \to FN$, cf. [2, p. 226]) be the natural lift of $f \in \mathrm{Diff}N$ to the bundle of metrics (resp. linear frame bundle); namely $f_{\mathscr{M}}(g_x) = (f^{-1})^* g_x$ (resp. $\tilde{f}(X_1, \dots, X_n) = (f_* X_1, \dots, f_* X_n)$, where $(X_1, \dots, X_n) \in F_x(N)$); hence $p_{\mathscr{M}} \circ f_{\mathscr{M}} = f \circ p_{\mathscr{M}}$ (resp. $p_F \circ \tilde{f} = f \circ p_F$), and $f_{\mathscr{M}} \colon \mathscr{M} \to \mathscr{M}$ (resp. $\tilde{f} \colon FN \to FN$) have a natural extension to jet bundles $f^{(r)}_{\mathscr{M}} \colon J^r(\mathscr{M}) \to J^r(\mathscr{M})$ (resp. $\tilde{f}^{(r)} \colon J^r(FN) \to J^r(FN)$) as defined in the formula (1), i.e.,

$$f^{(r)}_{\mathscr{M}}(j^r_x g) = j^r_{f(x)}(f_{\mathscr{M}} \circ g \circ f^{-1}) \quad (\text{resp.}\, \tilde{f}^{(r)}(j^r_x s) = j^r_{f(x)}(\tilde{f} \circ s \circ f^{-1})).$$

If f_t is the flow of a vector field $X \in \mathfrak{X}(N)$, then the infinitesimal generator of $(f_t)_M$ (resp. $\tilde{f}_t$) in $\mathrm{Diff}\mathscr{M}$ (resp. $\mathrm{Diff}FN$) is denoted by $X_{\mathscr{M}}$ (resp. $\tilde{X}$) and the following Lie-algebra homomorphisms are obtained:

$$\begin{cases} \mathfrak{X}(N) \to \mathfrak{X}(\mathscr{M}), \ X \mapsto X_{\mathscr{M}} \\ \mathfrak{X}(N) \to \mathfrak{X}(FN), \ X \mapsto \tilde{X}. \end{cases}$$

2.5 Diff$N$- and $\mathfrak{X}(N)$ -Invariance

A differential form $\omega_k \in \Omega^k(J^r(\mathcal{M}))$, $k \in \mathbb{N}$, is said to be Diff$N$-invariant—or invariant under diffeomorphisms—(resp. $\mathfrak{X}(N)$-invariant) if the following equation holds: $(f^{(r)}_{\mathcal{M}})^*\omega_k = \omega_k$, $\forall f \in \mathrm{Diff}N$ (resp. $L_{X^{(r)}_{\mathcal{M}}}\omega_k = 0$, $\forall X \in \mathfrak{X}(N)$). Obviously, Diff$N$-invariance implies $\mathfrak{X}(N)$-invariance and the converse is almost true (see [1, 4]). Because of this, below we consider $\mathfrak{X}(N)$-invariance only.

As N is an oriented manifold, there exists a unique p-horizontal n-form $\mathbf{v}_g$ on $\mathcal{M}$ such that, $\mathbf{v}_{g_x}(X_1, \dots, X_n) = 1$, for every g_x-orthonormal basis $(X_1, \dots, X_n)$ belonging to the orientation of N. Locally $\mathbf{v}_g = \rho\mathbf{v}$, where $\rho = \sqrt{(-1)^{n^-}\det(y_{ij})}$ and $\mathbf{v} = dx^1 \wedge \cdots \wedge dx^n$. As proved in [7, Proposition 7], the form $\mathbf{v}_g$ is Diff$N$-invariant and hence $\mathfrak{X}(N)$-invariant.

A Lagrangian density Λ on $J^r(\mathcal{M})$ can be globally written as $\Lambda = \mathcal{L}\mathbf{v}_g$ for a unique function $\mathcal{L} \in C^\infty(J^r(\mathcal{M}))$ and Λ is $\mathfrak{X}(N)$-invariant if and only if the function $\mathcal{L}$ is $\mathfrak{X}(N)$-invariant. Therefore, the invariance of Lagrangian densities is reduced to that of scalar functions.

A Lagrangian density Λ defined on $J^r(FN)$ can be written as follows: $\Lambda = \mathcal{L}\theta^1 \wedge \cdots \wedge \theta^n$, where $\theta = (\theta^1, \dots, \theta^n)$ is the canonical 1-form on FN ([2, III, Sect. 2, p. 118]), and $\mathcal{L} \in C^\infty(J^r(FN))$ is called the 'canonical Lagrangian' associated to Λ. A Lagrangian density Λ is Diff$N$-invariant (resp. $\mathfrak{X}(N)$-invariant) if and only if $\mathcal{L} \circ J^r(\tilde{\phi}) = \mathcal{L}$, $\forall \phi \in \mathrm{Diff}N$ (resp. $\tilde{X}^{(r)}(\mathcal{L}) = 0$, $\forall X \in \mathfrak{X}(N)$), as θ is Diff$N$-invariant and hence, $\mathfrak{X}(N)$-invariant. Diff$N$-invariance implies $\mathfrak{X}(N)$-invariance and both notions are equivalent except when N is orientable and admits an orientation-reversing diffeomorphism onto itself (see [4, Sect. 2.1]).

3 A Basis of $\mathcal{I}_{\mathfrak{X}(N)}$

Let $\mathcal{L}^i_{jk} \colon J^1(FN) \to \mathbb{R}$, $j < k$, be the Lagrangian $\mathcal{L}^i_{jk}(j^1_x s) = \omega^i([X_j, X_k])(x)$, where $s = (X_1, \dots, X_n)$ and $(\omega^1, \dots, \omega^n)$ denotes the dual coframe. We remark that the definition makes sense as the value $\omega^i([X_j, X_k])(x)$ only depends on $j^1_x s$. Moreover, from the very definition we have $[X_j, X_k]_x = \mathcal{L}^i_{jk}(j^1_x s)(X_i)_x$. The local expression of this Lagrangian in an induced coordinate system on $J^1(FN)$, is

$$\mathcal{L}^i_{jk} = (x^h_j x^l_{k,h} - x^h_k x^l_{j,h})z^i_l, \tag{3}$$

where $z^i_j = (x^i_j)^{-1}$. The Lagrangians $\mathcal{L}^i_{jk}$ are Diff$N$-invariant and functionally independent and every $\mathcal{L} \in \mathcal{I}_{\mathfrak{X}(N)}$ can be written locally as a differentiable function of this system (see [1]).

As is known (see [1, Theorem 4.8], [5, formula (6)]), every $\mathfrak{X}(N)$-invariant Lagrangian on $J^2(FN)$ can be written as a differentiable function of the following

$$\tfrac{1}{2}n^2(n-1) + \tfrac{1}{6}n^2(n-1)(2n+2)$$

Lagrangians: $\mathscr{L}^c_{ab}$, $a < b$; $\mathscr{L}^c_{ab,d}$, $a < b, a \leq d$, where $\mathscr{L}^c_{ab}$ is defined by the formula (3) and $\mathscr{L}^c_{ab,d}$ is given as follows:

$$\begin{aligned}\mathscr{L}^c_{ab,d} &= \sum_r x^r_d D_r \left(\mathscr{L}^c_{ab}\right) \\ &= x^r_d (x^u_{a,r} x^e_{b,u} - x^u_{b,r} x^e_{a,u}) z^c_e - x^r_d x^s_{t,r} (x^u_a x^e_{b,u} - x^u_b x^e_{a,u}) z^c_s z^t_e \\ &+ x^r_d (x^v_a x^s_{b,(rv)} - x^v_b x^s_{a,(rv)}) z^c_s, \end{aligned} \tag{4}$$

where D_i denotes the total derivative with respect to the coordinate x^i; that is,

$$D_i = \frac{\partial}{\partial x^i} + \sum_{|I|=0}^{\infty} x^h_{k,I+(i)} \frac{\partial}{\partial x^h_{k,I}},$$

$I = (i_1, \ldots, i_n) \in \mathbb{N}^n$ being a multi-index of order $|I| = i_1 + \cdots + i_n$.

The geometric meaning of such functions is the following. If $s \colon U \to FN$ is the section induced by a linear frame $(X_1, \ldots, X_n)$ on an open neighbourhood of a point $x \in N$, we denote by ∇^s the only linear connection on FU that parallelizes each vector field X_i; i.e., $(\nabla^s)_{X_j} X_i = 0$, for $i, j = 1, \ldots, n$. Then, we have

$$\begin{aligned}\mathscr{L}^c_{ab}(j^1_x s) &= -\omega^c \left(\mathrm{Tor}_{\nabla^s} (X_a, X_b)\right)(x), \\ \mathscr{L}^c_{ab,d}(j^2_x s) &= -\omega^c \left(\nabla^s \mathrm{Tor}_{\nabla^s}\right)(X_d, X_a, X_b)(x),\end{aligned}$$

where $(\omega^1, \ldots, \omega^n)$ denotes the dual coframe. Moreover, the inequality $a \leq d$ is due to the constraints imposed by Bianchi's first identity for the linear connection ∇^s; namely,

$$\mathscr{L}^h_{ab}\mathscr{L}^c_{hd} + \mathscr{L}^h_{bd}\mathscr{L}^c_{ha} + \mathscr{L}^h_{da}\mathscr{L}^c_{hb} = \mathscr{L}^c_{bd,a} + \mathscr{L}^c_{da,b} + \mathscr{L}^c_{ab,d}.$$

For the details of these facts, see [1, 4].

4 Einstein–Hilbert Lagrangian on $\mathscr{M}$

Following the notations in [2, Chap. VI, Sect. 5], the Ricci tensor field attached to the symmetric connection Γ is given by $S^\Gamma(X, Y) = \mathrm{trace}(Z \mapsto R^\Gamma(Z, X)Y)$, where R^Γ denotes the curvature tensor field of the covariant derivative ∇^Γ associated to Γ on the tangent bundle; hence $S^\Gamma = (R^\Gamma)_{jl} dx^l \otimes dx^j$, where $(R^\Gamma)_{jl} = (R^\Gamma)^k_{jkl}$, and

$$(R^\Gamma)^i_{jkl} = \frac{\partial \Gamma^i_{jl}}{\partial x^k} - \frac{\partial \Gamma^i_{jk}}{\partial x^l} + \Gamma^m_{jl}\Gamma^i_{km} - \Gamma^m_{jk}\Gamma^i_{lm}, \tag{5}$$

(see [2, Chap. III, Proposition 7.6]).

The Einstein–Hilbert Lagrangian density is given by

$$(\Lambda_{EH})_{j^2_x g} = g^{ij}(x)(R^g)^h_{ihj}(x)\mathbf{v}_g(x) = L_{EH}(j^2_x g)\mathbf{v}_x,$$

where $\mathbf{v} = dx^1 \wedge \cdots \wedge dx^n$, R^g is the curvature tensor of the Levi-Civita connection Γ of the metric g, and $\mathbf{v}_g$ denotes the Riemannian volume form attached to g; i.e., in coordinates, $\mathbf{v}_g = \sqrt{(-1)^{n^-} | \det((g_{ab})^n_{a,b=1})|}\mathbf{v}$. Hence,

$$L_{EH} \circ j^2 g = (\rho \circ g)(y^{ij} \circ g)(R^g)^h_{ihj}, \quad \rho = \sqrt{(-1)^{n^-} | \det((y_{ab})^n_{a,b=1})|},$$

where y_{ij} is introduced in (2) and $y^{ij} = (y_{ij})^{-1}$.

Taking (5) into account, the local expression for L_{EH} is readily seen to be

$$L_{EH} = \rho y^{ij}\left(\frac{\partial \Gamma^h_{ij}}{\partial x^h} - \frac{\partial \Gamma^h_{ih}}{\partial x^j} + \Gamma^k_{ij}\Gamma^h_{hk} - \Gamma^k_{ih}\Gamma^h_{jk}\right),$$

or, in terms of the local coordinates on $J^2(\mathcal{M})$,

$$\begin{aligned} L_{EH} &= \rho y^{ij} y^{hk}\left(y_{ki,(jh)} - y_{ij,(kh)}\right) \\ &\quad + \tfrac{1}{2}\rho y^{ij} y^{ak} y^{bh}\left(-y_{ab,h}\left(y_{ki,j} + y_{kj,i} - y_{ij,k}\right)\right. \\ &\qquad + y_{ab,j} y_{kh,i} + \tfrac{1}{2}\left(y_{ai,j} + y_{aj,i} - y_{ij,a}\right) y_{bh,k} \\ &\qquad \left. - \tfrac{1}{2}\left(y_{ai,h} + y_{ah,i} - y_{ih,a}\right)\left(y_{bk,j} + y_{bj,k} - y_{kj,b}\right)\right). \end{aligned} \tag{6}$$

5 The Einstein–Hilbert Lagrangian Induced on *FN*

The bundle of metrics $p_{\mathcal{M}}\colon \mathcal{M} \to N$ can be viewed as the quotient bundle $q\colon FN \to FN/G \cong \mathcal{M}$, where FN is the bundle of linear frames of N and $G = O(n^+, n^-) \subset GL(n; \mathbb{R})$ is the orthogonal group corresponding to the signature (n^+, n^-), $n^+ + n^- = n = \dim N$. In fact, if $(X_1, \ldots, X_n)$ is a linear frame at $x \in N$ with dual coframe $(\omega^1, \ldots, \omega^n)$, then the identification between FN/G and $\mathcal{M}$ is given by the bundle map

$$(X_1, \ldots, X_n) \mathrm{mod} G \mapsto g_x = (\omega^1)^2 + \cdots + (\omega^{n^+})^2 - (\omega^{n^+ + 1})^2 - \cdots - (\omega^n)^2.$$

A linear frame $(X_1, \ldots, X_n) \in F_x(N)$ is said to be orthonormal with respect to $g_x \in \mathcal{M}_x(N)$ (or simply g_x-orthonormal) if $g_x(X_i, X_j) = \varepsilon_i \delta_{ij}$ where $\varepsilon_i = 1$ for $1 \leq i \leq n^+$ and $\varepsilon_i = -1$ for $n^+ + 1 \leq i \leq n$.

Theorem 1 *The Lagrangian $\bar{L}_{EH}$ induced on $J^2(FN)$ by the Einstein–Hilbert Lagrangian L_{EH} can be written as a differentiable function of the basis of* Diff$N$*-invariant Lagrangians as follows:*

$$\bar{L}_{EH} = \bar{\rho}\varepsilon_r \left(2\mathscr{L}^a_{ra,r} - \mathscr{L}^a_{ar}\mathscr{L}^b_{br} - \tfrac{1}{2}\mathscr{L}^a_{rb}\left(\mathscr{L}^b_{ra} - \varepsilon_a\varepsilon_b\mathscr{L}^a_{rb}\right)\right),$$

where

$$\bar{\rho} = \sqrt{\det((z^a_b)^n_{a,b=1})}.$$

Proof Let $q^{(2)}\colon J^2(FN) \to J^2(\mathscr{M})$ be the morphism induced by the bundle morphism $q\colon FN \to \mathscr{M}$. Let $\bar{y}_{jk} = y_{jk} \circ q$, $\bar{y}_{ij,k} = y_{ij,k} \circ q^{(2)}$ and $\bar{y}_{ij,(kh)} = y_{ij,(kh)} \circ q^{(2)}$ be the coordinates system on $J^2(\mathscr{M})$ induced by the coordinates on $J^2(FN)$; that is,

$$\bar{y}_{jk} = \varepsilon_i z^i_j z^i_k, \tag{7}$$

$$\bar{y}_{ij,k} = -\varepsilon_a z^a_r \left(z^s_i z^a_j + z^a_i z^s_j\right) x^r_{s,k}, \tag{8}$$

$$\begin{aligned}\bar{y}_{ij,(kh)} = {} & \varepsilon_a z^a_u x^u_{v,h} z^v_r \left(z^s_i z^a_j + z^a_i z^s_j\right) x^r_{s,k} \\ & + \varepsilon_a z^a_r \left(z^s_u z^a_j + z^a_u z^s_j\right) z^v_i x^u_{v,h} x^r_{s,k} \\ & + \varepsilon_a z^a_r \left(z^s_i z^a_u + z^a_i z^s_u\right) z^v_j x^u_{v,h} x^r_{s,k} \\ & - \varepsilon_a z^a_r \left(z^s_i z^a_j + z^a_i z^s_j\right) x^r_{s,(kh)},\end{aligned} \tag{9}$$

where z^i_j is introduced in Sect. 3.

Taking (6) into account, the local expression for the Einstein–Hilbert Lagrangian induced on $J^2(FN)$ is written as follows:

$$\bar{L}_{EH} = \bar{\rho}\bar{y}^{ij}\bar{y}^{hk}\left(\bar{y}_{ki,(jh)} - \bar{y}_{ij,(kh)}\right) + \tfrac{1}{2}\bar{\rho}(\bar{L}_{EH})_0,$$

where

$$\begin{aligned}(\bar{L}_{EH})_0 = \bar{y}^{ij}\bar{y}^{ak}\bar{y}^{bh} \big(-\bar{y}_{ab,h}\left(\bar{y}_{ki,j} + \bar{y}_{kj,i} - \bar{y}_{ij,k}\right) \\ + \bar{y}_{ab,j}\bar{y}_{kh,i} + \tfrac{1}{2}\left(\bar{y}_{ai,j} + \bar{y}_{aj,i} - \bar{y}_{ij,a}\right)\bar{y}_{bh,k} \\ -\tfrac{1}{2}\left(\bar{y}_{ai,h} + \bar{y}_{ah,i} - \bar{y}_{ih,a}\right)\left(\bar{y}_{bk,j} + \bar{y}_{bj,k} - \bar{y}_{kj,b}\right)\big),\end{aligned}$$

and

$$\bar{\rho} = \sqrt{|\det((\bar{y}_{ab})^n_{a,b=1})|}.$$

Taking (7)–(9) and

$$\bar{y}^{rs} = \varepsilon_i x^r_i x^s_i,$$

into account, after long but simple calculations, we obtain

$$
\begin{aligned}
\bar{L}_{EH} &= 2\bar{\rho}\varepsilon_w x_w^k \left(x_w^h x_{a,(kh)}^r - x_a^h x_{w,(kh)}^r\right) z_r^a \\
&\quad + \bar{\rho}\left\{2\varepsilon_w x_w^h x_{w,h}^r x_{s,u}^u z_r^s - 4\varepsilon_w x_w^h x_w^k z_u^a x_{v,h}^u x_{a,k}^r z_r^v + 2\varepsilon_w x_w^h x_{w,r}^u x_{s,h}^r z_u^s\right. \\
&\qquad + 2\varepsilon_a\varepsilon_p\varepsilon_w x_w^h \left(x_p^k x_{w,k}^r - x_w^k x_{p,k}^r\right) z_r^a z_u^a x_{p,h}^u \\
&\qquad - \tfrac{1}{2}\varepsilon_r x_{r,b}^a x_{r,a}^b - \varepsilon_r x_{r,a}^a x_{r,b}^b \\
&\qquad - 2\varepsilon_r x_r^h x_{r,h}^b z_b^n x_{n,a}^a + 2\varepsilon_r x_r^h z_b^n x_{n,h}^b x_{r,a}^a + 2\varepsilon_r x_r^h z_b^n x_{r,h}^a x_{n,a}^b \\
&\qquad - \varepsilon_r x_r^h z_b^n x_{r,a}^b x_{n,h}^a - \varepsilon_r x_r^a x_r^b z_s^h z_u^k x_{k,b}^u x_{h,a}^s \\
&\qquad + \tfrac{3}{2}\varepsilon_r x_r^a x_r^b z_s^h z_u^k x_{h,b}^u x_{k,a}^s - \varepsilon_m\varepsilon_w\varepsilon_n z_r^n z_s^n x_m^b x_{m,b}^s x_w^a x_{w,a}^r \\
&\qquad \left.+ \tfrac{3}{2}\varepsilon_n\varepsilon_m\varepsilon_w z_s^n z_r^n x_m^a x_m^b x_{w,b}^r x_{w,a}^s - \tfrac{1}{2}\varepsilon_m\varepsilon_n\varepsilon_w z_t^n z_r^n x_m^a x_{w,a}^r x_w^b x_{m,b}^t\right\}.
\end{aligned}
$$

Finally, taking (3) and

$$
\mathscr{L}_{ma,m}^a = x_m^h x_{m,h}^u x_{v,u}^r z_r^v - x_m^h x_{v,h}^u z_u^a x_m^k x_{a,k}^r z_r^v + x_m^h (x_m^k x_{a,(hk)}^r - x_a^k x_{m,(hk)}^r) z_r^a,
$$

(see (4)) into account we obtain the statement.

Remark 1 The Einstein–Hilbert Lagrangian $\bar{L}_{EH}$ induced on $J^2(FN)$ is an affine function and its Poincaré–Cartan form projects onto $J^1(FN)$ (see [6, Proposition 2.1]).

Let τ_N^1, τ_N^2 be the mappings given by,

$$
\begin{aligned}
\tau_N^1 &\colon J^1(FN) \longrightarrow \wedge^2 T^*N \otimes TN, \quad \tau_N^1(j_x^1 s) = (\mathrm{Tor}_{\nabla^s})_x\,, \\
\tau_N^2 &\colon J^2(FN) \longrightarrow T^*N \otimes \wedge^2 T^*N \otimes TN, \quad \tau_N^2(j_x^2 s) = (\nabla^s \mathrm{Tor}_{\nabla^s})_x\,,
\end{aligned}
$$

where ∇^s is the linear connection parallelizing the linear frame bundle defined by the section s.

Let $\bar{C}_3^1$ and C_1^1 be the contractions given by

$$
\begin{aligned}
\bar{C}_3^1 &\colon T^*N \otimes \wedge^2 T^*N \otimes TN \longrightarrow T^*N \otimes T^*N, \\
\bar{C}_3^1 &\left(\tau_{abc}^d dx^a \otimes dx^b \wedge dx^c \otimes \tfrac{\partial}{\partial x^d}\right) = \tau_{abc}^c dx^a \otimes dx^b,
\end{aligned}
$$

$$
\begin{aligned}
C_1^1 &\colon \wedge^2 T^*N \otimes TN \longrightarrow T^*N, \\
C_1^1 &\left(\tau_{ab}^c dx^a \wedge dx^b \otimes \tfrac{\partial}{\partial x^c}\right) = \tau_{cb}^c dx^b,
\end{aligned}
$$

and let $\tilde{C}_4^1$, $\tilde{C}_2^2$ be the contractions given by

$$\tilde{C}_4^1, \tilde{C}_2^2 \colon \wedge^2 T^*N \otimes TN \otimes \wedge^2 T^*N \otimes TN \longrightarrow T^*N \otimes T^*N,$$
$$\tilde{C}_4^1 \left(\tau_{aba'b'}^{cc'} dx^a \wedge dx^b \otimes \frac{\partial}{\partial x^c} \otimes dx^{a'} \wedge dx^{b'} \otimes \frac{\partial}{\partial x^{c'}} \right) = \tau_{aba'c}^{cc'} dx^a \wedge dx^b \otimes dx^{a'} \otimes \frac{\partial}{\partial x^{c'}},$$
$$\tilde{C}_2^2 \left(\tau_{aba'b'}^{cc'} dx^a \wedge dx^b \otimes \frac{\partial}{\partial x^c} \otimes dx^{a'} \wedge dx^{b'} \otimes \frac{\partial}{\partial x^{c'}} \right) = \tau_{ac'a'b'}^{cc'} dx^a \otimes dx^{a'} \wedge dx^{b'} \otimes \frac{\partial}{\partial x^c}.$$

Proposition 1 *The Einstein–Hilbert functional induced on $J^2(FN)$ is the second-order Lagrangian density $\bar{\mathscr{L}}_{EH}\theta^1 \wedge \cdots \wedge \theta^n$ on FN where $\bar{\mathscr{L}}_{EH}$ is defined as follows*

$$\bar{\mathscr{L}}_{EH} = g_s^* \circ \left(2\left(\tilde{C}_3^1 \circ \tau_N^2\right) - \left(\left(C_2^1 \circ \tau_N^1\right) \otimes \left(C_2^1 \circ \tau_N^1\right)\right) - \tfrac{1}{2}\left(\left(\tilde{C}_4^1 \otimes \tilde{C}_2^2\right) \circ \left(\tau_N^1 \otimes \tau_N^1\right)\right)\right)$$
$$+ \tfrac{1}{2}\left(g_s^\sharp \circ \left(\tau_N^1 \otimes \tau_N^1\right)\right),$$

where $g_s^* \colon T^*N \otimes T^*N \longrightarrow \mathbb{R}$, $g_s^* = (X_1)^2 + \cdots + (X_{n^+})^2 - (X_{n^++1})^2 - \ldots - (X_n)^2$, with $s = (X_1, \ldots, X_n)$, and $g_s^\sharp$ is the map induced by g_s^* and g_s on $\wedge^2 T^*N \otimes TN \otimes \wedge^2 T^*N \otimes TN$.

References

1. García, P.L., Muñoz Masqué, J.: Differential invariants on the bundles of linear frames. J. Geom. Phys. **7**, 395–418 (1990)
2. Kobayashi, S., Nomizu, K.: Foundations of Differential Geometry, vol. I. Wiley, New York (1963)
3. Luehr, C.P., Rosenbaum, M.: Gravitation as an internal gauge theory of the Poincaré group. J. Math. Phys. **21**, 1432–1438 (1980)
4. Muñoz Masqué, J., Rosado, M.E.: Diff*N* does not act transitively on the solutions of an invariant variational problem on *FN*, differential geometry and applications. In: Janyška, J., Kolař, I., Slovák, J. (eds.) Proceedings of 6th International Conference on DGA at Brno, pp. 161–169. Masaryk University, Brno, Czech Republic (1996)
5. Muñoz Masqué, J., Rosado, M.E.: Invariant variational problems on linear frame bundles. J. Phys. A **35**(8), 2013–2036 (2002)
6. Muñoz Masqué, J., Rosado, M.E.: Integrability of second-order Lagrangians admitting a first-order Hamiltonian formalism. Differ. Geom. Appl. **35**, 164–177 (2014)
7. Muñoz Masqué, J., Valdés Morales, A.: The number of functionally independent invariants of a pseudo-Riemannian metric. J. Phys. A: Math. Gen. **27**, 7843–7855 (1994)
8. Ne'eman, Y.: Gravity is the gauge theory of the parallel-transport. Modification of the Poincaré Group. Lecture Notes in Mathematics, vol. 676, pp. 189–215. Springer, Berlin (1977)
9. van Elst, H., Uggla, C.: General relativistic 1+3 orthonormal frame approach. Class. Quantum Gravity **14**, 2673–2695 (1997)

Printed by Printforce, the Netherlands